Geophysical Monograph Series

Including

IUGG Volumes

Maurice Ewing Volumes

Mineral Physics Volumes

GEOPHYSICAL MONOGRAPH SERIES

Geophysical Monograph Volumes

1. Antarctica in the International Geophysical Year *A. P. Crary, L. M. Gould, E. O. Hulburt, Hugh Odishaw, and Waldo E. Smith (Eds.)*
2. Geophysics and the IGY *Hugh Odishaw and Stanley Ruttenberg (Eds.)*
3. Atmospheric Chemistry of Chlorine and Sulfur Compounds *James P. Lodge, Jr. (Ed.)*
4. Contemporary Geodesy *Charles A. Whitten and Kenneth H. Drummond (Eds.)*
5. Physics of Precipitation *Helmut Weickmann (Ed.)*
6. The Crust of the Pacific Basin *Gordon A. Macdonald and Hisashi Kuno (Eds.)*
7. Antarctica Research: The Matthew Fontaine Maury Memorial Symposium *H. Wexler, M. J. Rubin, and J. E. Caskey, Jr. (Eds.)*
8. Terrestrial Heat Flow *William H. K. Lee (Ed.)*
9. Gravity Anomalies: Unsurveyed Areas *Hyman Orlin (Ed.)*
10. The Earth Beneath the Continents: A Volume of Geophysical Studies in Honor of Merle A. Tuve *John S. Steinhart and T. Jefferson Smith (Eds.)*
11. Isotope Techniques in the Hydrologic Cycle *Glenn E. Stout (Ed.)*
12. The Crust and Upper Mantle of the Pacific Area *Leon Knopoff, Charles L. Drake, and Pembroke J. Hart (Eds.)*
13. The Earth's Crust and Upper Mantle *Pembroke J. Hart (Ed.)*
14. The Structure and Physical Properties of the Earth's Crust *John G. Heacock (Ed.)*
15. The Use of Artificial Satellites for Geodesy *Soren W. Henricksen, Armando Mancini, and Bernard H. Chovitz (Eds.)*
16. Flow and Fracture of Rocks *H. C. Heard, I. Y. Borg, N. L. Carter, and C. B. Raleigh (Eds.)*
17. Man-Made Lakes: Their Problems and Environmental Effects *William C. Ackermann, Gilbert F. White, and E. B. Worthington (Eds.)*
18. The Upper Atmosphere in Motion: A Selection of Papers With Annotation *C. O. Hines and Colleagues*
19. The Geophysics of the Pacific Ocean Basin and Its Margin: A Volume in Honor of George P. Woollard *George H. Sutton, Murli H. Manghnani, and Ralph Moberly (Eds.)*
20. The Earth's Crust: Its Nature and Physical Properties *John C. Heacock (Ed.)*
21. Quantitative Modeling of Magnetospheric Processes *W. P. Olson (Ed.)*
22. Derivation, Meaning, and Use of Geomagnetic Indices *P. N. Mayaud*
23. The Tectonic and Geologic Evolution of Southeast Asian Seas and Islands *Dennis E. Hayes (Ed.)*
24. Mechanical Behavior of Crustal Rocks: The Handin Volume *N. L. Carter, M. Friedman, J. M. Logan, and D. W. Stearns (Eds.)*
25. Physics of Auroral Arc Formation *S.-I. Akasofu and J. R. Kan (Eds.)*
26. Heterogeneous Atmospheric Chemistry *David R. Schryer (Ed.)*
27. The Tectonic and Geologic Evolution of Southeast Asian Seas and Islands: Part 2 *Dennis E. Hayes (Ed.)*
28. Magnetospheric Currents *Thomas A. Potemra (Ed.)*
29. Climate Processes and Climate Sensitivity (Maurice Ewing Volume 5) *James E. Hansen and Taro Takahashi (Eds.)*
30. Magnetic Reconnection in Space and Laboratory Plasmas *Edward W. Hones, Jr. (Ed.)*
31. Point Defects in Minerals (Mineral Physics Volume 1) *Robert N. Schock (Ed.)*
32. The Carbon Cycle and Atmospheric CO_2: Natural Variations Archean to Present *E. T. Sundquist and W. S. Broecker (Eds.)*
33. Greenland Ice Core: Geophysics, Geochemistry, and the Environment *C. C. Langway, Jr., H. Oeschger, and W. Dansgaard (Eds.)*
34. Collisionless Shocks in the Heliosphere: A Tutorial Review *Robert G. Stone and Bruce T. Tsurutani (Eds.)*
35. Collisionless Shocks in the Heliosphere: Reviews of Current Research *Bruce T. Tsurutani and Robert G. Stone (Eds.)*
36. Mineral and Rock Deformation: Laboratory Studies —The Paterson Volume *B. E. Hobbs and H. C. Heard (Eds.)*
37. Earthquake Source Mechanics (Maurice Ewing Volume 6) *Shamita Das, John Boatwright, and Christopher H. Scholz (Eds.)*
38. Ion Acceleration in the Magnetosphere and Ionosphere *Tom Chang (Ed.)*
39. High Pressure Research in Mineral Physics (Mineral Physics Volume 2) *Murli H. Manghnani and Yasuhiko Syono (Eds.)*
40. Gondwana Six: Structure Tectonics, and Geophysics *Gary D. McKenzie (Ed.)*
41. Gondwana Six: Stratigraphy, Sedimentology, and Paleontology *Garry D. McKenzie (Ed.)*

42 **Flow and Transport Through Unsaturated Fractured Rock**   *Daniel D. Evans and Thomas J. Nicholson (Eds.)*

43 **Seamounts, Islands, and Atolls**   *Barbara H. Keating, Patricia Fryer, Rodey Batiza, and George W. Boehlert (Eds.)*

44 **Modeling Magnetospheric Plasma**   *T. E. Moore and J. H. Waite, Jr. (Eds.)*

45 **Perovskite: A Structure of Great Interest to Geophysics and Materials Science**   *Alexandra Navrotsky and Donald J. Weidner (Eds.)*

46 **Structure and Dynamics of Earth's Deep Interior (IUGG Volume 1)**   *D. E. Smylie and Raymond Hide (Eds.)*

47 **Hydrological Regimes and Their Subsurface Thermal Effects (IUGG Volume 2)**   *Alan E. Beck, Grant Garven, and Lajos Stegena (Eds.)*

48 **Origin and Evolution of Sedimentary Basins and Their Energy and Mineral Resources (IUGG Volume 3)**   *Raymond A. Price (Ed.)*

49 **Slow Deformation and Transmission of Stress in the Earth (IUGG Volume 4)**   *Steven C. Cohen and Petr Vaníček (Eds.)*

50 **Deep Structure and Past Kinematics of Accreted Terranes (IUGG Volume 5)**   *John W. Hillhouse (Ed.)*

51 **Properties and Processes of Earth's Lower Crust (IUGG Volume 6)**   *Robert F. Mereu, Stephan Mueller, and David M. Fountain (Eds.)*

52 **Understanding Climate Change (IUGG Volume 7)**   *Andre L. Berger, Robert E. Dickinson, and J. Kidson (Eds.)*

53 **Plasma Waves and Instabilities at Comets and in Magnetospheres**   *Bruce T. Tsurutani and Hiroshi Oya (Eds.)*

54 **Solar System Plasma Physics**   *J. H. Waite, Jr., J. L. Burch, and R. L. Moore (Eds.)*

55 **Aspects of Climate Variability in the Pacific and Western Americas**   *David H. Peterson (Ed.)*

56 **The Brittle-Ductile Transition in Rocks**   *A. G. Duba, W. B. Durham, J. W. Handin, and H. F. Wang (Eds.)*

57 **Evolution of Mid Ocean Ridges (IUGG Volume 8)**   *John M. Sinton (Ed.)*

58 **Physics of Magnetic Flux Ropes**   *C. T. Russell, E. R. Priest, and L. C. Lee (Eds.)*

59 **Variations in Earth Rotation (IUGG Volume 9)**   *Dennis D. McCarthy and Williams E. Carter (Eds.)*

60 **Quo Vadimus Geophysics for the Next Generation (IUGG Volume 10)**   *George D. Garland and John R. Apel (Eds.)*

61 **Cometary Plasma Processes**   *Alan D. Johnstone (Ed.)*

62 **Modeling Magnetospheric Plasma Processes**   *Gordon K. Wilson (Ed.)*

63 **Marine Particles Analysis and Characterization**   *David C. Hurd and Derek W. Spencer (Eds.)*

64 **Magnetospheric Substorms**   *Joseph R. Kan, Thomas A. Potemra, Susumu Kokubun, and Takesi Iijima (Eds.)*

65 **Explosion Source Phenomenology**   *Steven R. Taylor, Howard J. Patton, and Paul G. Richards (Eds.)*

66 **Venus and Mars: Atmospheres, Ionospheres, and Solar Wind Interactions**   *Janet G. Luhmann, Mariella Tatrallyay, and Robert O. Pepin (Eds.)*

67 **High-Pressure Research: Application to Earth and Planetary Sciences (Mineral Physics Volume 3)**   *Yasuhiko Syono and Murli H. Manghnani (Eds.)*

68 **Microwave Remote Sensing of Sea Ice**   *Frank Carsey, Roger Barry, Josefino Comiso, D. Andrew Rothrock, Robert Shuchman, W. Terry Tucker, Wilford Weeks, and Dale Winebrenner*

69 **Sea Level Changes: Determination and Effects (IUGG Volume 11)**   *P. L. Woodworth, D. T. Pugh, J. G. DeRonde, R. G. Warrick, and J. Hannah*

70 **Synthesis of Results from Scientific Drilling in the Indian Ocean**   *Robert A. Duncan, David K. Rea, Robert B. Kidd, Ulrich von Rad, and Jeffrey K. Weissel (Eds.)*

71 **Mantle Flow and Melt Generation at Mid-Ocean Ridges**   *Jason Phipps Morgan, Donna K. Blackman, and John M. Sinton (Eds.)*

72 **Dynamics of Earth's Deep Interior and Earth Rotation (IUGG Volume 12)**   *Jean-Louis Le Mouël, D.E. Smylie, and Thomas Herring (Eds.)*

73 **Environmental Effects on Spacecraft Positioning and Trajectories (IUGG Volume 13)**   *A. Vallance Jones (Ed.)*

74 **Evolution of the Earth and Planets (IUGG Volume 14)**   *E. Takahashi, Raymond Jeanloz, and David Rubie (Eds.)*

75 **Interactions Between Global Climate Subsystems: The Legacy of Hann (IUGG Volume 15)**   *G. A. McBean and M. Hantel (Eds.)*

76 **Relating Geophysical Structures and Processes: The Jeffreys Volume (IUGG Volume 16)**   *K. Aki and R. Dmowska (Eds.)*

77 **The Mesozoic Pacific: Geology, Tectonics and Volcanism—A Volume in Memory of Sy Schlanger**   *Malcolm S. Pringle, William W. Sager, William V. Sliter, and Seth Stein (Eds.)*

78 **Climate Change in Continental Isotopic Records**   *P. K. Swart, K. C. Lohmann, J. McKenzie, and S. Savin (Eds.)*

79 **The Tornado: Its Structure, Dynamics, Prediction, and Hazards**   *C. Church, D. Burgess, C. Doswell, R. Davies-Jones (Eds.)*

80 **Auroral Plasma Dynamics**   *R. L. Lysak (Ed.)*

81 **Solar Wind Sources of Magnetospheric Ultra-Low Frequency Waves**   *M. J. Engebretson, K. Takahashi, and M. Scholer (Eds.)*

82 **Gravimetry and Space Techniques Applied to Geodynamics and Ocean Dynamics** *Bob E. Schutz, Allen Anderson, Claude Froidevaux, and Michael Parke (Eds.)*

Maurice Ewing Volumes

1 **Island Arcs, Deep Sea Trenches, and Back-Arc Basins** *Manik Talwani and Walter C. Pitman III (Eds.)*
2 **Deep Drilling Results in the Atlantic Ocean: Ocean Crust** *Manik Talwani, Christopher G. Harrison, and Dennis E. Hayes (Eds.)*
3 **Deep Drilling Results in the Atlantic Ocean: Continental Margins and Paleoenvironment** *Manik Talwani, William Hay, and William B. F. Ryan (Eds.)*
4 **Earthquake Prediction—An International Review** *David W. Simpson and Paul G. Richards (Eds.)*
5 **Climate Processes and Climate Sensitivity** *James E. Hansen and Taro Takahashi (Eds.)*
6 **Earthquake Source Mechanics** *Shamita Das, John Boatwright, and Christopher H. Scholz (Eds.)*

IUGG Volumes

1 **Structure and Dynamics of Earth's Deep Interior** *D. E. Smylie and Raymond Hide (Eds.)*
2 **Hydrological Regimes and Their Subsurface Thermal Effects** *Alan E. Beck, Grant Garven, and Lajos Stegena (Eds.)*
3 **Origin and Evolution of Sedimentary Basins and Their Energy and Mineral Resources** *Raymond A. Price (Ed.)*
4 **Slow Deformation and Transmission of Stress in the Earth** *Steven C. Cohen and Petr Vaníček (Eds.)*
5 **Deep Structure and Past Kinematics of Accreted Terrances** *John W. Hillhouse (Ed.)*
6 **Properties and Processes of Earth's Lower Crust** *Robert F. Mereu, Stephan Mueller, and David M. Fountain (Eds.)*
7 **Understanding Climate Change** *Andre L. Berger, Robert E. Dickinson, and J. Kidson (Eds.)*
8 **Evolution of Mid Ocean Ridges** *John M. Sinton (Ed.)*
9 **Variations in Earth Rotation** *Dennis D. McCarthy and William E. Carter (Eds.)*
10 **Quo Vadimus Geophysics for the Next Generation** *George D. Garland and John R. Apel (Eds.)*
11 **Sea Level Changes: Determinations and Effects** *Philip L. Woodworth, David T. Pugh, John G. DeRonde, Richard G. Warrick, and John Hannah (Eds.)*
12 **Dynamics of Earth's Deep Interior and Earth Rotation** *Jean-Louis Le Mouël, D.E. Smylie, and Thomas Herring (Eds.)*
13 **Environmental Effects on Spacecraft Positioning and Trajectories** *A. Vallance Jones (Ed.)*
14 **Evolution of the Earth and Planets** *E. Takahashi, Raymond Jeanloz, and David Rubie (Eds.)*
15 **Interactions Between Global Climate Subsystems: The Legacy of Hann** *G. A. McBean and M. Hantel (Eds.)*
16 **Relating Geophysical Structures and Processes: The Jeffreys Volume** *K. Aki and R. Dmowska (Eds.)*
17 **Gravimetry and Space Techniques Applied to Geodynamics and Ocean Dynamics** *Bob E. Schutz, Allen Anderson, Claude Froidevaux, and Michael Parke (Eds.)*

Mineral Physics Volumes

1 **Point Defects in Minerals** *Robert N. Schock (Ed.)*
2 **High Pressure Research in Mineral Physics** *Murli H. Manghnani and Yasuhiko Syona (Eds.)*
3 **High Pressure Research: Application to Earth and Planetary Sciences** *Yasuhiko Syono and Murli H. Manghnani (Eds.)*

Geophysical Monograph 83
IUGG Volume 18

Nonlinear Dynamics and Predictability of Geophysical Phenomena

William I. Newman
Andrei Gabrielov
Donald L. Turcotte

Editors

American Geophysical Union
International Union of Geodesy and Geophysics

Published under the aegis of the AGU Books Board

Library of Congress Cataloging-in-Publication Data

Nonlinear dynamics and predictability of geophysical phenomena /
   William I. Newman, Andrei Gabrielov, Donald L. Turcotte, editors.
      p. cm. — (Geophysical monograph ; 83) (IUGG ; v. 18)
   Includes bibliographical references.
   ISBN 0-87590-469-6
   1. Geodynamics.   2. Chaotic behavior in systems.   3. Nonlinear
theories.   I. Newman, William I.   II. Gabrielov, A. M.
III. Turcotte, Donald Lawson.   IV. Series.   V. Series: IUGG (Series)
; v. 18.
QE501.3.N66   1994
550—dc20   94-20388
                                                                         CIP

ISSN: 0065-8448
ISBN 0-87590-469-6

Copyright 1994 by the International Union of Geodesy and Geophysics and the American Geophysical Union, 2000 Florida Avenue, NW, Washington, DC 20009, U.S.A.

Figures, tables, and short excerpts may be reprinted in scientific books and journals if the source is properly cited.

   Authorization to photocopy items for internal or personal use, or the internal or personal use of specific clients, is granted by the American Geophysical Union for libraries and other users registered with the Copyright Clearance Center (CCC) Transactional Reporting Service, provided that the base fee of $1.00 per copy plus $0.10 per page is paid directly to CCC, 222 Rosewood Dr., Danvers, MA 01923. 0065-8448/94/$01.00+0.20.
   This consent does not extend to other kinds of copying, such as copying for creating new collective works or for resale. The reproduction of multiple copies and the use of full articles or the use of extracts, including figures and tables, for commercial purposes requires permission from AGU.

Printed in the United States of America.

CONTENTS

Preface
William I. Newman, Andrei Gabrielov, Donald L. Turcotte   ix

Foreword
Helmut Moritz   xi

1. **Chaos: A Historical Perspective**
   Sir James Lighthill   1

2. **Seismicity Modeling and Earthquake Prediction: A Review**
   Andrei Gabrielov and William I. Newman   7

3. **Nonlinear Dynamics of the Transition from Stable Sliding to Cyclic Stick-Slip in Rock**
   Yaojun Gu and Teng-fong Wong   15

4. **Is the Dynamics of the Lithosphere Chaotic?**
   Q. Li and E. Nyland   37

5. **Dynamics of a Seismic Regime: Vrancea—A Case History**
   Cezar-Ioan Trifu and Mircea Radulian   43

6. **The Precursor of Instability for Nonlinear Systems and Its Application to Earthquake Prediction—the Load-Unload Response Ratio Theory**
   Xiang-chu Yin, Can Yin, and Xue-zhong Chen   55

7. **Strange Attractor in Nonlinear Fluctuations of Length of the Day (LOD) Time Series**
   R. K. Tiwari, J. G. Negi, and K. N. N. Rao   61

8. **Self-Organized Criticality: Consequences for Statistics and Predictability of Earthquakes**
   Per Bak, Kim Christensen, and Zeev Olami   69

9. **Period-Doubling Bifurcation and Chaotic Phenomena in a Single Degree of Freedom Elastic System with a Two-State Variable Friction Law**
   Niu Zhiren and Chen Dangmin   75

10. **Nonlinear Dynamic Modeling of Earthquake Prediction**
    Yaolin Shi, Lumin Geng, and Guomin Zhang   81

11. **Methods for Improving the Prediction of Dynamical Processes with Special Reference to the Atmospheric Circulation**
    Johan Grasman and Peter Houtekamer   91

12. **The Nonlinear Asymptotic Stage of the Rayleigh-Taylor Instability with Wide Bubbles and Narrowing Spikes**
    V. M. Cherniavski and Yu. M. Shtemler   97

13. **Correlation Dimension of the Strange Attractor for Geomagnetic Field Variations**
    Yu. S. Tyupkin and A. Ya. Feldstein   103

PREFACE

The goal of this volume is to establish an understanding and interdisciplinary cooperation among geophysicists and nonlinear dynamicists. While the last thirty years has brought substantial progress in the study of the atmosphere and ocean as well as of convection in the Earth's mantle and core, the nonlinear revolution is only beginning to have an impact on the investigation of the solid Earth. The problem of predictability in chaotic nonlinear systems is one of the most important and difficult subjects in modern nonlinear science. In its application to geophysics and, especially, earthquake prediction, it presents both a profound intellectual problem and an issue with important societal implications.

The contributions to this volume have been organized according to intellectual areas and methodological approaches. Most of the papers deal with the solid Earth, while others focus on fluids and plasmas. The papers range from those dealing with experimental and observational treatments of real materials and systems and their analysis, often by modern nonlinear techniques, to the theoretical and computational modeling of different geophysical systems. Indeed, this volume encompasses not only many disciplines but also spatial and temporal domains that extend over many orders of magnitude.

An outline of the historical development of the problem of predictability is presented in the first of the articles in this volume, namely "Chaos: A Historical Perspective," by Sir James Lighthill. While this article focuses almost exclusively on problems having an underlying Hamiltonian character, the assemblage of problems confronted by geophysics are rarely Hamiltonian and display an even richer array of phenomena.

Gabrielov and Newman then shift the focus to the solid Earth environment and review seismicity modeling and earthquake prediction: this article provides a bridge between the various disciplines that have shed light upon the earthquake prediction problem. Given the overall importance of rheology and material properties, this section continues with an article by Gu and Wong, in which they investigate in the laboratory the nonlinear dynamics of the transition from stable sliding to cyclic stick-slip behavior in rocks.

The next set of articles present data which demonstrate the existence of nonlinearity in the solid Earth. Li and Nyland employ catalogued seismic data for western Canada to show the presence of low-dimensional chaos. Trifu and Radulian explore the intermediate-depth seismic regime in time, space, and energy for Vrancea, Romania, suggesting possible precursors. Yin et al. explore the possibility of using loading rates as an empirical precursor for Chinese seismicity and employ this property to predict a number of earthquakes. The last of the papers to investigate observational data is devoted to more global issues: Tiwari et al. identify a strange attractor and nonlinear fluctuations in the length of day.

The development of theoretical models in geophysics has provided vital new insights into nonlinear phenomena and, especially, earthquake prediction. Bak et al. develop a self-organized criticality (SOC) model for earthquake events and explore the consequences of its statistics and the predictability of earthquakes. While SOC has become somewhat controversial in the nonlinear community, Zhiren and Dangmin consider instead a one degree of freedom elastic system with a two-state friction law and show that it can be described by a familiar Feigenbaum sequence en route to chaos. Finally, Shi et al. also consider a classical set of models based on block-spring-dashpot design and show their relevance to the temporal, spatial, and magnitude distribution of seismicity.

In the final part of this volume, nonlinearity in fluid environments is considered. Grasman and Houtekamer present a variety of methods for assessing and reducing errors in establishing the initial state of a nonlinear system, methods whose utility extends far beyond the atmospheric context considered by the authors. Cherniavsky and Shtemler consider the role of the Rayleigh-Taylor instability, and the potential for numerical pathologies, in the simulation of the Rayleigh-Taylor instability's effect on mantle convection. Finally, Tyupkin and Feldstein consider the correlation dimension of the strange attractor and geomagnetic field variations, evidence for chaotic behavior in the magnetosphere.

This volume provides a representative set of papers (as well as some invited contributions) based on the Union

Symposium "Nonlinear Dynamics and Predictability of Critical Geophysical Phenomena," part of the twentieth General Assembly of the International Union of Geodesy and Geophysics, held in Vienna, August 1991, which help define the role that nonlinearity and predictability have come to occupy in the Earth sciences. The production of this volume faced many challenges. In addition to the nonlinear revolution that is sweeping much of contemporary science, another revolution--the revitalization of Eastern Europe--added to the challenge. This volume is extraordinarily diverse in both its intellectual and international content.

The editors hope that this volume will help establish some of the themes for further investigation into nonlinearity and, especially, predictability in the geophysical sciences. The Symposium which was the stimulus for this volume generated some conclusions that have broad applicability to the general subject and may provide important insight into the selection of future research topics.

First, the application of methods of nonlinear science provides a powerful tool for the study of predictability both in the atmospheric and solid Earth sciences. Chaotic behavior of nonlinear systems does not exclude predictability, but introduces an upper bound on the ability to make predictions and renders predictions probabilistic. Indeed, an important question which emerges is that of establishing the "predictability of predictability"-the estimation of the time-scale over which forecasts are reliable given the current state of the system.

Second, a common feature of many atmospheric and solid Earth phenomena is an essentially hierarchical structure. This, in turn, may impose a bound on predictability depending on the scale of spatio-temporal averaging. Thus, different types of methods are required for making predictions at different scales within the hierarchy. Unfortunately, this also argues for the lack of reliable prediction during certain stages, such as intermediate-term forecasting in the atmospheric sciences and short-term prediction in the solid Earth sciences.

Third, the key ingredient in prediction is an adequate physical model. The major success enjoyed in atmospheric science is largely due to the development of general circulation models. However, because of the multitude of interacting mechanisms and the absence of fundamental constitutive equations, there remains no adequate general model of the processes in the lithosphere. In the Earth sciences, the principal theoretical advances are connected with the detailed study of nonlinear rheology of fault zones and with the understanding of the role of hierarchical self-organization due to the interaction of lithospheric blocks. While the practical outcome of work in earthquake prediction remains empirically based, the ideas of nonlinear science are employed in the search for precursory phenomena and for assessing the reliability of such methods.

We hope that the insights gained from this volume will serve as a source of new research problems for geophysicists and their nonlinear dynamicist colleagues.

William I. Newman
University of California
Los Angeles, California

Andrei Gabrielov
University of Toronto
Toronto, Canada

Donald L. Turcotte
Cornell University
Ithaca, New York

Editors

FOREWORD

The scientific work of the International Union of Geodesy and Geophysics (IUGG) is primarily carried out through its seven associations: IAG (briefly, Geodesy), IASPEI (Seismology), IAVCEI (Volcanology), IAGA (Geomagnetism), IAMAP (Meteorology), IAPSO (Oceanography), and IAHS (Hydrology). The work of these associations is documented in various ways.

Nonlinear Dynamics and Predictability of Geophysical Phenomena is one of a group of volumes published jointly by IUGG and AGU that are based on work presented at the Inter-Association Symposia as part of the IUGG General Assembly held in Vienna, Austria, in August 1991. Each symposium was organized by several of IUGG's member associations and comprised topics of interdisciplinary relevance. The subject areas of the symposia were chosen such that they would be of wide interest. Also, the speakers were selected accordingly, and in many cases, invited papers of review character were solicited. The series of symposia were designed to give a picture of contemporary geophysical activity, results, and problems to scientists having a general interest in geodesy and geophysics.

In view of the importance of these interdisciplinary symposia, IUGG is grateful to AGU for having put its unique resources in geophysical publishing expertise and experience at the disposal of IUGG. This ensures accurate editorial work, including the use of peer reviewing. So the reader can expect to find expertly published scientific material of general interest and general relevance.

Helmut Moritz
President, IUGG

Chaos: A Historical Perspective

Sir James Lighthill

Department of Mathematics, University College London, London

1. SEVENTEENTH-CENTURY BACKGROUND

In this introductory lecture I'd like to offer a broad historical perspective regarding the *relatively* recent general recognition:

(a) that mechanical systems satisfying Newton's laws may be subject to the essentially unpredictable type of behavior which the word CHAOS describes--in other words, the recognition

(b) that quantum effects are not required;

(c) so that, notwithstanding Heisenberg, uncertainty is there on the basis of the good old classical mechanics based on Newton's Laws.

But first of all I'll remind you that there are two kinds of laws in science, which we may exemplify by Kepler's Laws and Newton's Laws.

Kepler in 1609 completed some very detailed observations of the motions of Mars; together with a full geometrical description of them, in the Copernican sun-centered frame of reference, as motions in a constant orbit in the shape of an ellipse with the Sun as focus. A decade later Kepler had published the *Epitome Astronomiae Copernicanae* (a rather more substantial work than the *Dialogo* which later got Galileo into some difficulties), and had there described in detail his most famous discovery: Kepler's three empirical laws concerning planetary orbits. These laws, of the elliptical shapes of orbits, of the radius covering equal areas in equal times, and of the proportionality of the square of the orbital period to the cube of the major axis, were shown from the observations to be closely satisfied by the Earth and by the five then known planets; and furthermore, by the four satellites of Jupiter which Galileo had recently discovered.

In beginning a lecture on Newtonian dynamics I have thought it appropriate to highlight some of those 'shoulders of giants' on which Newton recognized himself as standing; and, in particular, to recall the massive tasks of accumulating observational data, and of determining empirical laws that would describe them accurately, which made possible Newton's achievements. Nevertheless, I want to make a big distinction between any empirically based laws like those of Kepler, which may approximately summarize some large mass of data in a convenient and (perhaps) philosophically intriguing form, and physico-mathematical laws embodied in a system of general applicability such as Newton was the first to introduce.

The modern world's practitioners of Mechanics, for whom I do in some sense have the right to speak because I have just completed four years as President of the International Union of Theoretical and Applied Mechanics, feel pride in the knowledge that Mechanics was the first science for which a systematic framework of physico-mathematical laws was constructed, being expounded (of course) in the *Principia* which Newton completed just over three hundred years ago, in 1687.

And from these laws I'll here pick out Newton's inverse-square law of gravitation--which was called "universal" not just because he showed it to be the only law of force compatible with the three empirical laws of Kepler regarding motions of planets around the Sun (and of satellites around planets); but also because the same law accounted for gravitational phenomena at the surface of the Earth. In the case of an attracting body with spherical symmetry, the effect of attraction by all the particles making it up was proved equivalent to attraction by the whole mass concentrated at the center; in such a way that another body moving solely in that field of force would describe an elliptic orbit with that center as focus, and in accordance also with the other two laws of Kepler.

Newton recognized, at the same time, how dynamical effects of the Earth's rotation about its axis, taken in combination with his law of gravitation, accounted for the observed spheroidal shape of the Earth's surface, flattened at the poles. He appreciated, furthermore, how this departure from spherical symmetry made a certain *perturbation* in the gravitational force acting on the Moon which would partly explain the tendency of its orbit to precess.

Post-Newtonian analyses of the solar system have been much concerned with developing such perturbation theories in far greater detail, with the object in particular of understanding all of those departures from the original simple empirical laws of Kepler which later astronomers were able to uncover in the course of their still more accurate observations. A particularly famous perturbation theory has been the three-body problem, concerned with how the orbit of one body around another is perturbed by the influence of a third body. Many of the most important of the later developments in Newtonian mechanics (which, I should perhaps remind you, is the mechanics used for almost all practical engineering purposes) have been founded on mathematical models first introduced in the context of perturbation theory; methods to which I must now briefly refer.

2. MATHEMATICAL STUDIES OF NEWTONIAN DYNAMICAL SYSTEMS

(a) Developments in Perturbation Theory

Some big procedural advances in mathematical analysis, and in particular the theory of differential equations, were needed for these subsequent major developments in Newtonian dynamics. Work done in the 18th century by the great Swiss mathematician Leonhard Euler initiated these advances, which were continued by Joseph Louis Lagrange. Between them they developed a method for calculating how different types of perturbations would effect a slow change in an orbit's so-called *elements*. These are a set of quantities defining the geometry of an orbital ellipse, and the orientation of its plane in space; quantities which under Kepler's laws would remain constant but which can vary slowly in response to perturbing forces of all kinds. Modern space scientists apply this method devised by Euler and Lagrange to relate observed variations in the orbital elements of artificial Earth satellites to the different types of perturbation causing them, and so to obtain precise estimates of strengths of different components in the perturbing forces; and this method gives us the best currently available values, in particular, for the departures of the Earth's real surface shape from the simple spheroidal first approximation obtained by Newton.

In the meantime, Lagrange himself had taken matters much farther in his *magnum opus,* the *Mécanique Analytique,* to obtain an absolutely general form for the system of differential equations satisfied by any mechanical system obeying Newton's laws. His own life's work, and that of his younger but equally gifted colleague Pierre Simon de Laplace, combined to yield clear explanations in terms of Newton's laws for practically all the peculiarities of orbital phenomena observed in the solar system. Laplace expounded this at length in his five-volume *magnum opus,* the *Mécanique Céleste,* and accompanied it with the brilliantly written single volume *Exposition du Système du Monde* (1796) which popularized the successes of Newtonian dynamics in accounting both for the intricate behavior of the solar system and for (broadly speaking) its stability. This work must have done much to spread into the general consciousness a belief in the complete predictability of systems based on Newton's laws of motion; a belief, as it were, in the determinism of the mechanical universe.

Here I have to pause, and to speak once again on behalf of the broad global fraternity of practitioners of mechanics. We are all deeply conscious today that the enthusiasm of our forebears for the marvelous achievements of Newtonian mechanics led them to make generalizations in this area of predictability which, indeed, we may have generally tended to believe before 1960, but which we now recognize were false. We collectively wish to apologize for having misled the general educated public by spreading ideas about the determinism of systems satisfying Newton's laws of motion that, after 1960, were proved incorrect. In this lecture, I am trying to make belated amends by explaining both the very different picture that we now discern, and the reasons for it having been uncovered so late.

The flowering of rigorous mathematical analysis in the nineteenth century, particularly among the French school of analysts, led to a situation where difficult questions about (say) the properties of solutions to a system of differential equations might at last be settled with complete certainty through the mathematical proofs of appropriate theorems. Now the equations of a very general dynamical system, given (as we have seen) in one quite useful form by Lagrange, had been transformed in the eighteen-thirties by the brilliant Irish mathematician Sir William Rowan Hamilton into the still more convenient form of a set of simultaneous first-order differential equations; what we call the Hamiltonian formulation. Such a set of equations lent itself to study by rigorous mathematical analysis, which was attempted in particular at the turn of the century by Henri

Poincaré (not to be confused with the distinguished statesman of the same surname). Poincaré (1892) was especially concerned with difficult cases of the three-body problem including some of the interacting complexities of the Sun-Earth-Moon system which had continued to be hard to calculate.

Exact numerical computations of the solutions to such a Hamiltonian system of equations could be attempted, of course, only much later, in the post-1960 period, when the availability of powerful enough computers was certainly one of the main factors that altered our view of how such solutions behave. For Poincaré and for all of the pre-1960 analysts it was necessary to base their analysis on perturbation theory; beginning, that is, from a very simple first approximation such as an orbit satisfying Kepler's laws, and then proceeding by taking into account perturbing effects (for example, of a third body) to a higher approximation, and then to successively higher and higher approximations, and finally seeking to prove rigorously that the procedure converged to a well-determined limit. It was a program that could be carried out very effectively in a wide range of cases. Nevertheless, in another important set of cases where there existed possibilities of approximate resonance between multiples of the different orbital periods of two interacting oscillations in a system, Poincaré's method of proof failed because of a difficulty that became known as the difficulty of small divisors, or resonant divisors. These were the denominators of terms in an infinite series in which every so often one of the terms had a very small denominator and this could often prevent the series from converging.

It was easy, of course, to assume that the failure of Poincaré's method of proof in a certain range of cases was accidental and that a regrouping of terms in the series, a sort of renormalization procedure such as achieved great success in other physical theories, would rectify the proofs so that they could be made rigorous. That type of approach was attempted for many years but it could never be shown to be applicable in full generality, and we now realize that this was due to a fundamentally new type of behavior, which occurs in certain dynamical systems satisfying Newton's laws; a type which we call 'chaotic'. The rest of this lecture must be devoted to indicating how chaotic behavior was discovered and to describing its nature (see Lichtenberg and Lieberman (1983) for a general survey).

(b) Chaotic Behavior

Three different new directions of research initiated the discovery of chaotic behavior in simple systems satisfying Newton's laws during the early part of the 1960s; although, admittedly, it was only twenty years later that the remarkably widespread occurrence of chaotic behavior had become clear; to such an extent that, for example, at the 1984 Congress of the International Union of Theoretical and Applied Mechanics the specialized topic which was highlighted most strongly in the Congress program was that of chaotic behavior.

The first of these three new directions of research arose directly out of Poincaré's work on nonlinear perturbation theory, related to equations in Hamiltonian form for isolated systems of constant energy. Work initiated by the great Russian analyst Kolmogorov (1954), and pursued by his gifted colleague, Arnold (1963), had been aimed first of all towards filling in the gaps in Poincaré's proofs; and, indeed, all this work along with independent studies by Moser (1962) in Germany and America demonstrated that, even in the neighborhood of resonances, perturbations did assume a distinctly regular pattern in the vast majority of cases. Nevertheless, gaps in that regular pattern did exist; namely, very small ranges of initial conditions for which the motion assumed a form described as chaotic or stochastic (that is, random). It was regarded as interesting that equations of motion that included absolutely no random element should possess solutions which behaved in such a random way. Initially, however, the matter was seen as something of a curiosity because the highly complex proofs used in the perturbation theory were valid only when the perturbations were restricted to being sufficiently small and, in this case, the ranges of initial conditions for which chaotic or stochastic motions occurred were very limited indeed.

The second new direction of research utilized the powerful computers that were by then available to compute solutions not just in cases when the perturbations were small, but also for much larger perturbations. This work, carried out, for example, by Greene (1979) in the U.S.A. and by Chirikov (1979) in the U.S.S.R., demonstrated that as the strength of the perturbations continued to grow there was a sharp increase in the range of initial conditions for which solutions behaved stochastically. Finally, at a certain level of the perturbation strength, the authors observed what they called a transition to global stochasticity, with all solutions behaving chaotically. I shall describe later what this amounts to in detail, but in the meantime will note the obvious fact that the new data made a big increase in the importance to be attached to chaotic or stochastic solutions.

These results on isolated systems of constant energy were of interest not only to astronomy but also to thermodynamics. The second law of thermodynamics envisages, of course, an increase in the randomness of motions experienced by an isolated system of molecules;

that is, an increase in its entropy; but physicists had long supposed that large numbers of collisions between molecules were necessary to allow such randomization to occur. Now, with the wider understanding of how chaotic motions can develop, it is possible to see that collisions may not be essential. For example, the ionized gas between the Sun and the Earth with its extremely low density, producing an astronomically large mean free path between molecular collisions, may nevertheless in the presence of magnetic fields experience phenomena that are possible only with an increase in entropy. One of these, which spacecraft have observed, is the so-called 'bow shock wave' where the solar wind of charged particles emanating from the sun is abruptly slowed down where it first encounters the Earth's magnetosphere.

But that is a digression, which may on the other hand have provided a valuable reminder that Newtonian dynamics is applicable not only to systems of solid bodies but also to fluid systems, including ordinary gases and liquids. Ordinary gases and liquids, of course, are subject to the phenomenon called viscosity, which causes the mechanical energy in their shearing motions to be gradually dissipated into heat; precisely as a result of an entropy increase associated with normal molecular collisions on a submicroscopic scale. Yet even the damping of fluid motions by viscosity does not prevent perfectly regular fluid motions from becoming chaotic and this fact was first made precise over a century ago, in 1883, by Osborne Reynolds. He showed how the regular flow of fluid through a pipe suddenly become chaotic or turbulent when the force producing the motion becomes sufficiently large relative to the damping forces due to viscosity. He showed, furthermore, that this randomization has nothing to do with the random molecular movements occurring at submicroscopic scales; they, indeed, have a damping effect tending to reduce the trend towards turbulent motions; motions which themselves involve, rather, a chaotic pattern of fluid movement on a strictly macroscopic scale.

Thus, the specialists in dynamics of fluids, such as myself, have long been most fully conscious of the common tendency for regular or laminar motions of fluids to become chaotic or turbulent even though the motions in question are subject to energy dissipation by the action of viscosity. On the other hand, fluids represent very complicated dynamical systems with an essentially unlimited number of degrees of freedom (each separate particle of fluid is separately free to be arbitrarily positioned relatively to all the other particles) and it had never been clear whether or not this was an essential prerequisite for chaotic behavior to develop.

Against that background it may be interesting to note that the third new direction of research which began in the early nineteen-sixties was concerned with some quite simple systems analogous to turbulence. These were dynamical systems with energy dissipation and just two or three degrees of freedom which, although forced in a perfectly regular way, responded in a completely chaotic way when a ratio of forcing effects to damping effects (a ratio similar to the Reynolds number introduced by Reynolds) was sufficiently large. Initially, they were devised by some noted experts in dynamics of fluids, including the great atmospheric scientist E. N. Lorenz, in order to mimic as closely as possible the development of turbulence in fluid systems. Lorenz (1963) introduced the term 'strange attractor' to describe the type of randomized motion which inexorably tends to develop.

More recently, a very general theory of these strange attractors has been produced for such systems, which unlike the isolated energy-preserving systems studied by Poincaré and others, are subjected both to forcing and to damping (see chapter 7 of Lichtenberg and Lieberman 1983). This theory suggests the steps by which regular motions develop into chaotic motions as some forcing: damping ratio changes. Often that takes place via an infinite sequence of so-called 'period-doubling bifurcations' which terminate, after just a finite change in that ratio, in a completely chaotic motion. When an infinite sequence of bifurcations has occurred, they produce beyond a certain time horizon that completely randomized dependence on initial conditions which destroys predictability. Numerical computations of very high precision have excellently confirmed these theories and demonstrated the strong tendency for systems of this type also to develop chaotic motions.

And various more complicated systems about which you'll be hearing today exhibit still more complicated forms of chaos than these. But in every case certain common facts relevant to predictability persist (Lighthill 1986).

3. LIMITS TO PREDICTABILITY FOR CHAOTIC SYSTEMS IN GENERAL

Here I remind you that, when a system is well defined, predictability depends upon the sensitivity of system behavior to initial conditions. More precisely it is concerned with whether, given the initial conditions to whatever is the appropriate number of decimal places, the system's future may be approximately forecast.

Now, the common feature that is found to be characteristic of all chaotic systems is the existence of a predictability horizon. This is a time after which solutions with initial conditions that are 'nearest neighbors' to the accuracy of specification being used here become remote from one another; doing so, furthermore, in a manner which varies in a discontinuous and randomized way in response to changes in just the last decimal place being used to describe the initial conditions.

You will want to ask, of course, what happens if we insist on *more* decimal places being used to describe the initial conditions. The answer is that predictability is changed rather little. In fact, as the number of decimal places increases, the predictability horizon changes with it only *linearly*. This is because chaotic systems exhibit the property that neighboring solutions diverge exponentially from one another (see, for example, p. 262 *et seq.* of Lichtenberg and Liebermann 1983).

So we see that, whatever level of accuracy can be achieved in the measurement of the initial state, alternative possibilities for the future become quite randomized (so that prediction is impossible) beyond a certain time horizon: the predictability horizon. Now I pointed out earlier that within systems governed by Newtonian dynamics it has been established that a fair proportion of them exhibit the chaotic property, and that this proportion increases as a function of the perturbation strength. For these systems, there is only a finite predictability horizon if initial conditions are prescribed to a given number of decimal places.

Now it may, of course, happen that some of you will wish to question the relevance of this conclusion by pointing out that, if the initial conditions were prescribed to an infinite number of decimal places, then the solution would be uniquely determined. In response to that, a mathematician might want in the first place to query whether the mathematician's own concept of the so-called 'real number' defined by an infinite sequence of decimal places has any relevance at all to measurements of what we habitually call 'the real world'. However, in the context of Newtonian dynamics we do not need to speculate about this.

Necessarily, we use Newtonian dynamics to describe the macroscopic behavior of matter. This, however, can never be specified to more than a certain level of accuracy because of the well known Brownian motions associated with the continued vibrations of all the individual molecules; and, incidentally, we now recognize those vibrations of molecular systems as having the type of random behavior which they are known to exhibit (a type of random behavior usually called 'ergodic') just because of the propensity towards chaos in mechanical systems satisfying Newton's laws. Initial conditions can never be specified, then, beyond a certain level of precision, and this implies a predictability horizon.

This, then, was a historical account of how it became recognized that systems subject to the laws of Newtonian dynamics include a substantial proportion of systems that are chaotic; and that, for these latter systems, there is no predictability beyond a finite predictability horizon. We are able to come to this conclusion without ever having to mention quantum mechanics or Heisenberg's uncertainty principle. A fundamental uncertainty about the future is there, indeed, even on the supposedly solid basis of the good old laws of motion of Newton, which effectively *are* the laws of motion satisfied by macroscopic systems (that is, systems for which phenomena on the extreme submicroscopic scales specified by Planck's constant are negligible). I have ventured to feel that this historical account would be of interest to a meeting devoted to modern knowledge on chaos.

REFERENCES

Arnold, V. I., Proof of a theorem of A. N. Kolmogorov on the invariance of quasi-periodic motions under small perturbations of the Hamiltonian, *Usp. Math. Nauk*, 18, 13-40, translated as *Russian Math. Surveys*, vol. 18, no. 4, pp. 9-36, 1963.

Chirikov, B. V., A universal instability of many-dimensional oscillator systems, *Phys. Rep.*, 52, 263-379, 1979.

Greene, J. M., A method for determining a stochastic transition, *J. Math. Phys.*, 20, 1183-1201, 1979.

Kolmogorov, A. N., On the conservation of quasi-periodic motions for a small change in the Hamiltonian function, (In Russian), *Dokl. Akad. Nauk*, 98, 527-530, 1954.

Lichtenberg, A. J., and M. A. Lieberman, *Regular and Stochastic Motion*, Springer-Verlag, New York, 1983.

Lighthill, J., The recently recognised failure of predictability in Newtonian dynamics, Proc. Roy. Soc., A 407, 35-50, 1986.

Lorenz, E. N., Deterministic nonperiodic flow, *J. Atmos. Sci.*, 20, 130-141, 1963.

Moser, J., On invariant curves of area-preserving mappings of an annulus, *Nachr. Akad. Wiss. Göttingen, Math. Phys. Kl*, 2, 1-20, 1962.

Poincaré, H., *Les Methodes Nouvelles de la Mécanique Céleste*, 3 vols., Gauthier-Villars, Paris, 1892.

Sir James Lighthill, Department of Mathematics, University College London, Gower Street, London WC1E 6BT, United Kingdom

Seismicity Modeling and Earthquake Prediction: A Review

ANDREI GABRIELOV

Department of Geological Sciences and Mathematical Sciences Institute, Cornell University, Ithaca, NY 14853; and International Institute for Earthquake Prediction Theory and Mathematical Geophysics, Russian Academy of Sciences, Moscow

WILLIAM I. NEWMAN

Departments of Earth and Space Sciences, Astronomy, and Mathematics, University of California, Los Angeles, CA 90024

The modeling of seismicity, i.e. of earthquake sequences in contrast with individual events, is an important component of the earthquake prediction problem. Observations provide a hint toward a physical basis for premonitory patterns and suggest the existence of precursory phenomena. The possibility of producing sufficiently long model "catalogs" would permit the testing of statistical significance. Moreover, it would be possible to adjust the numerical parameterization of premonitory patterns found in advance of the few strong earthquakes that have been observed. Recently, methods and ideas of nonlinear science found applications to models of seismicity and to methods of earthquake prediction. We present here a brief review of different approaches to modeling of seismicity.

INTRODUCTION

One of the principal directions in earthquake prediction studies is the search for precursors of strong earthquakes based on anomalous patterns observed in moderate seismicity. The principal source of observational data are "catalogs," which are listings of earthquake events according to their size (i.e. magnitude or energy), and the time and location of occurrence. (For comparison purposes, "catalogs" are also produced from numerical models of earthquakes.) Patterns, then, have a decidedly statistical and often ephemeral meaning. Numerous seismicity patterns have been suggested as precursory phenomena [Mogi, 1985; Keilis-Borok, 1990a; Ma et al., 1990; Bonin, 1991]. This includes different types of activation, quiescence—including combinations of activation and quiescence—anomalous aftershock and foreshock activity, seismic migration, and spatio-temporal concentration of moderate earthquakes before a strong earthquake.

Unfortunately, none of these phenomena taken individually are reliable indicators, and few of them have been tested in forward prediction. Moreover, in many seismically active regions, reliable seismicity catalogs with sufficient numbers and kinds of seismicity patterns for the identification of precursory phenomena are available for too short a time period, thus excluding any statistical tests. Therefore, real progress in this area is impossible without an adequate model of the seismotectonic process, one that simulates both the occurrence and interaction of strong and moderate earthquakes.

The following are among the principal features of the lithosphere that should be incorporated into a model for it to be regarded as adequate:

- Interaction of the processes of different physical origin, and of different spatial and temporal scales;
- Hierarchical block or possibly "fractal" structure; and
- Self-similarity in space, time, and energy.

The traditional approach to modeling is based on one *specific* tectonic fault and, often, one strong earthquake in order to reproduce certain pre- and/or post-seismic phenomena (relevant to this specific earthquake). In contrast, the recently developed class of the slider-block and cellular automata models treats the seismotectonic process in the most abstract way, in order to reproduce *general* universal properties of seismicity, first of all, the Gutenberg-Richter frequency of occurrence law, starting from a homogeneous lattice of simple threshold elements.

The specific and general approaches have their respective advantages and disadvantages. The first approach, which takes into account detailed information on the local geotectonic environment, usually misses universal properties of a series of events in a system of interacting faults. The second approach may be considered to be a zero-order approximation to reality. However, the importance of this approach and, in general, the importance of the application of the methods of theoretical physics and nonlinear science to the earthquake prediction problem lies in the possibility of establishing generic analogs with problems in other sciences, and to elaborate a new language for the description of seismicity patterns on the basis of the well-developed lexicon of nonlinear science.

The presence of a large number of different nonlinear mechanisms relevant to the seismotectonic process suggests the applicability of the general approach of nonlinear science to complex dissipative systems [Keilis-Borok, 1990b], unveiling the universal patterns due to self-organization rather than investigating the numerous details of the specific mechanisms involved.

It seems, therefore, that an adequate model of seismicity should incorporate the universal features of self-organized nonlinear systems, as well as the specific geometry of interacting tectonic faults. In the following sections, we will review some of the most important features in modeling seismicity and earthquake prediction and go on to discuss their overall significance.

Earthquake Sequences

Earthquake sequences in real catalogs manifest some general features despite the different tectonic structures and levels of seismicity found in various seismic regions.

The sequence of earthquakes is apparently stationary; no noticeable trend has been discerned in the level of seismicity during the 100 or so years of detailed studies of world-wide seismicity. There is, in addition, a considerable stochastic component in the earthquake sequence.

Against this stationary stochastic background different regular patterns appear. The best known of these is the Gutenberg-Richter frequency of occurrence law

$$\log N(M) = \alpha - \beta M \tag{1}$$

where $N(M)$ is the distribution function of earthquakes above magnitude M [Gutenberg and Richter, 1944; Ishimoto and Iida, 1939]. This can be interpreted as an indicator of self-similarity of seismicity in magnitude (energy). The constant β is usually close to 1, which roughly corresponds to the uniform distribution of the total source area of earthquakes over a wide range of the source sizes. It should be noted, however, that the Gutenberg-Richter law is valid only within a certain range of magnitudes, with the lower cutoff at the magnitude below 3 [Aki, 1987] and the upper cutoff (depending on the seismic region) close to magnitude 9 for world-wide seismicity. There is also a noticeable break in this law around the magnitude 6–6.5, separating the weaker intra-crustal earthquakes from the larger earthquakes that rupture the entire seismogenic zone [Scholz, 1990, Ch. 4]. Also, the Gutenberg-Richter law is not applicable to the strongest earthquakes on a single fault, due to the existence of *characteristic earthquakes* with the magnitude related to the fault geometry [Schwartz et al., 1981; Schwartz and Coppersmith, 1984].

Other types of self-similarity include the Omori law [Omori, 1895] for the temporal distribution (self-similarity in time) of the number $n(t)$ of aftershocks of a strong earthquake

$$n(t) = \frac{c}{(1+t)^p} , \tag{2}$$

where p is near 1, and the fractal spatial distribution of the epicenters (self-similarity in space). See Kagan and Knopoff (1980), Kagan (1991), Takayasu (1990, p. 31), and Turcotte (1992, Ch. 4) for a discussion of these different empirical scaling laws.

Another type of regular behavior is the *migration* of earthquakes along tectonic structures [Mogi, 1968; Anderson, 1975; Lehner et al., 1981]. A useful way to think of this is that during some interval of time, say ten years, the majority of earthquakes in a given tectonic region are localized to some subset of this region. Then, in the following interval of time, the geographic center of earthquake activity has seemingly moved, giving rise to an apparent systematic migration.

Earthquakes often appear in *clusters*—see for example Rice and Gu (1983)—both in time and in space. One common clustering pattern is a main shock followed by a series of aftershocks. Foreshock activity is not so clearly expressed as aftershock sequences since the number of foreshocks of strong earthquakes is usually small or zero while aftershock series of such earthquakes can contain hundreds of events. Finally, there are doublets of strong earthquakes and "swarms" of earthquakes that cannot be separated into main shock, foreshocks and aftershocks because they have similar magnitudes.

The concept of the *seismic cycle*, originating with Reid (1910), implies characteristic time intervals between the strongest earthquakes in a certain region, with periods of post-seismic relaxation and inter-seismic stress accumulation between them—see Scholz (1990, Ch. 5) for a review. However, the actual time intervals are not equal and can deviate considerably from the average characteristic period which can vary from tens to hundreds of years. One of the most famous examples of a comparatively short seismic cycle is the Parkfield area of California where moderate earthquakes have recurred approximately every 22 years [Bakun and McEvily, 1984]. Nevertheless, the earthquake that according to this periodicity was expected around 1988, never occurred.

Numerous premonitory seismicity patterns have been found in earthquake catalogs. These include different types of activation—the general increase of activity, "swarms" of earthquakes, bursts of aftershocks, see Keilis-Borok et al. (1980)—and of quiescence or "seismic gap" [Wyss and Habermann, 1988; Ogata, 1992]. None of these premonitory patterns is reliable enough by itself and different combinations have been suggested for intermediate-term earthquake prediction. One particular combination of quiescence, in the area of a future strong earthquake and activation, in the surrounding areas, is called the "doughnut pattern" [Mogi, 1969]. More complicated combinations were suggested by Keilis-Borok and Rotwain (1990) and Keilis-Borok and Kossobokov (1990) based on pattern recognition methods. These premonitory patterns appear with a characteristic time scale of 1–5 years (intermediate-term prediction) and a characteristic space scale of hundreds of kilometers.

The two principal mechanisms involved in the seismotectonic process are tectonic loading, with characteristic rate of a few cm/yr, and the elastic stress accumulation and redistribution, with characteristic rate of $\approx$ 1 km/sec. In the typical time scale (10-100 yrs) of earthquake prediction studies, the first of these mechanisms can be considered to be a uniform rate of motion, and the second to be an instantaneous stress drop.

At the same time, there are several nonlinear mechanisms of the different physical nature that develop in the time scales intermediate between the two extremes, overlapping with the time scales of the premonitory seismicity patterns. This includes the spatial heterogeneity and hierarchical block structure of the lithosphere, different types of nonlinear rheology of the fault zones and friction along the fault planes, gravitational and thermodynamic processes, physical-chemical and phase transitions, fluid migration and stress corrosion. It is quite possible that these (and maybe some other, still unknown) mechanisms are responsible for the premonitory seismicity patterns characterized by the intermediate time scale and long-range spatial correlations in moderate seismicity preceding the strong earthquakes.

ELASTIC REBOUND THEORY

One important ingredient in the modeling of seismicity is based on the so-called elastic rebound theory [Reid, 1910] which emerged in the aftermath of the great San Francisco earthquake of 1906. According to this theory, elastic stress in a seismically active region accumulates due to some external source, e.g. movement of tectonic plates, and is released when the stress exceeds the strength of the medium. In the simplest case (constant rate of stress accumulation, fixed strength and residual stress) this model produces a periodic sequence of earthquakes of equal magnitude. This links the elastic rebound theory with the concepts of the seismic cycle and of characteristic earthquakes.

If only strength or residual stress is fixed in this model, we have the so-called "time-predictable" model (the time interval until the next earthquake is defined by the magnitude of the previous one) and the "slip-predictable" model (the magnitude of an expected earthquake increases with the elapsed time). Although a model of this type is used for long-term prediction [Nishenko and Buland, 1987], real sequences of strong earthquakes are fundamentally more complicated [Thatcher, 1990; Scholz, 1990, Ch. 5]. In particular, the elastic rebound model suggests that a strong earthquake should be followed by a period of quiescence, whereas in reality a strong earthquake is followed by a period of activation and sometimes by another earthquake of comparable magnitude. Simple deterministic nonlinear models for repetitive seismicity containing some of the attributes of "chaos" were developed by Newman and Knopoff (1982a,b) and by Knopoff and Newman (1983).

To incorporate the post-seismic activity following a strong earthquake, Elsasser (1969) suggested an additional viscous mechanism due to the interaction of the elastic upper crust with the asthenosphere and upper mantle. Yet another way to incorporate post-seismic activity is to include viscous interaction into rheology of the fault plane. A three-dimensional model of this type, with inhomogeneity in the distribution of the model parameters along the fault plane, was investigated by Mikumo and Miyatake (1983). These concepts were further developed to include the Maxwellian viscoelastic rheology as well as horizontal inhomogeneity [Rice, 1980; Lehner et al., 1981; Li and Rice, 1987; Ben-Zion et al., 1993)]. The aseismic (creeping) part in these models satisfies the constitutive equation

$$\sigma_{ij} = (K - \frac{2}{3}\mu)\epsilon_{kk}\delta_{ij} + 2\mu(\epsilon_{ij} - \epsilon_{ij}^{cr}) \quad , \tag{3}$$

where K and μ are elastic moduli, and the total strain ϵ is represented as a sum of the elastic strain ϵ^{el} and the inelastic creep strain ϵ^{cr}, i.e. $\epsilon_{ij} = \epsilon_{ij}^{el} + \epsilon_{ij}^{cr}$. In particular, Ben-Zion et al. (1993) consider the Parkfield sequence as a part of the great 1857 earthquake cycle and argue that, due to the relaxation mechanism, the frequency and magnitude of the earthquake in the sequence should actually decrease in time. Note that the presence of a threshold for failure introduces strong nonlinearity into these otherwise linear models.

RATE-DEPENDENT AND STATE-DEPENDENT FRICTION

A model with a rate-dependent and state-dependent friction law, based on laboratory experiments using rock samples, was introduced by Dieterich (1972) and further developed and studied by Ruina (1983), Tse and Rice (1986), and others. The model defines the dependence of the friction coefficient μ ($\tau = \mu\sigma$, where τ and σ are the tangent and normal stress components) on the slip velocity V and state variable θ according to

$$\mu = \mu_0 + a \ln\left(\frac{V}{V^*}\right) + b\theta \quad , \tag{4}$$

$$\frac{d\theta}{dt} = -\frac{V}{L}\left[\theta + \ln\left(\frac{V}{V^*}\right)\right] \quad . \tag{5}$$

Here V^*, μ_0, a, b, L are constants. For $a > b$ ("velocity strengthening") the model always gives stable sliding, and, for $a < b$ ("velocity weakening"), instability appears when the stiffness is below a critical value—namely, $-(a - b)\sigma/L$—see Gu et al. (1984). The model gives an adequate description of preseismic, coseismic and postseismic slip on a fault, especially when, as in Tse and Rice (1986), transition from velocity weakening to velocity strengthening with depth is included. See also Rice (1993) where the slip is allowed to vary along the strike, as well as with the depth, and an additional viscous damping term is added to account for the seismic radiation. Rice and Gu (1983) suggested that this friction law, together with relaxation processes in the lower lithosphere and asthenosphere, could be a possible mechanism for post-seismic activation effects. Lorenzetti and Tullis (1989) discussed possible implications of this model to short-term prediction based on preseismic slip measurements. Marone et al. (1991) suggest an opposite depth distribution of $a - b$ in the friction law (i.e. strengthening in the upper 3–5 km and weakening in the seismogenic layer) in order to explain earthquake afterslip at faults with a thick sedimentary cover.

The principal problem in this modeling is the applicability of the complicated friction law, derived from laboratory experiments on flat surfaces of homogeneous rock samples, to real fault zones that are neither homogeneous nor flat. The parameters in this friction law are empirical, and it is not clear how to scale them properly for real faults. The behavior of the system with this friction law is very sensitive to small variation in the values of the parameters—in the presence of noise, they may become virtually unpredictable.

SPATIAL HETEROGENEITY

Another direction in the modeling of complex earthquake sequences takes into account the spatial inhomogeneity of the strength distribution in the fault plane. The key concepts here are barriers, asperities, and characteristic earthquakes [Aki, 1984]. Asperities and barriers represent strong patches in the fault plane, while the difference is in their relation to the earthquake source. *Asperities* are strong patches on the stress-free background (due to preslip and foreshocks) and break during the earthquake [Kanamori and Stewart, 1978)]. *Barriers* appear as strong patches that do not allow further propagation of a fracture [Das and Aki, 1977; Aki, 1979]. The interpretation of barriers in terms of the geometry of tectonic faults was suggested by King and Nabelek (1985) and by King (1986). In particular, King (1986) suggested the existence of "soft" barriers where a seismic rupture terminates due to the absence of accumulated stress. Both asperities and barriers suggest the possible recurrence of earthquakes with a preferred source size, i.e. *characteristic earthquakes* [Schwartz et al., 1981].

STRESS CORROSION

Stress corrosion, or static fatigue [Anderson and Grew, 1977; Atkinson, 1984] is often considered to be one of the possible mechanisms for the time delay in the seismotectonic process. In this mechanism, especially in the presence of active fluids [Rhebinder and Shtchukin, 1972], the fractures in the stressed material grow and propagate quasi-statically under stresses that are substantially below the brittle fracture threshold and the effective strength of the material can be reduced by several orders of magnitude. This mechanism was suggested in Das and Scholz (1981), Newman and Knopoff (1982a, b), Knopoff and Newman (1983), and Yamashita and Knopoff (1987) to explain aftershock sequences. In Yamashita and Knopoff (1989) and Sornette et al. (1992), the stress corrosion mechanism was included in a model of foreshock activity. Gabrielov and Keilis-Borok (1983) considered geometrical patterns of the stress corrosion in inhomogeneously stressed medium as a possible mechanism for the spatial inhomogeneity of strength responsible for precursory phenomena, such as the doughnut pattern.

SLIDER-BLOCK MODELS AND SELF-ORGANIZED CRITICALITY

In contrast with the aforementioned models, a number of models composed of "masses and springs" or of cellular automata suggest the possibility of apparently chaotic earthquake sequences with a power law distribution of sizes in a spatially homogeneous medium due to self-organizing processes in a system of interacting elements (blocks, faults, etc.). The first class of these models, the slider-block models originally proposed by Burridge and Knopoff (1967), have been studied by Cao and Aki (1986), Takayasu and Matsuzaki (1988), Carlson and Langer (1989), Carlson et al. (1991), and others. In these models a linear system of rigid blocks connected by springs to adjacent blocks and to a driving slab and interacting with a stable surface according to a specified friction law.

In the original paper by Burridge and Knopoff, the model was shown to reproduce such important properties of seismicity as the Gutenberg-Richter law and, with the inclusion of additional viscous elements, aftershock activity. Cao and Aki (1986) considered a system of blocks with a rate-dependent and state-dependent friction law in order to reproduce premonitory quiescence. Carlson and Langer (1989) found a bimodal population of earthquakes in their model. While the small earthquakes obey a power law distribution, the strongest (runaway) events appear much more often than the extrapolation of the power law established for the small earthquakes would suggest. They associated this phenomenon with the concept of characteristic earthquakes. Shaw et al. (1992) reproduced activation and concentration patterns for small events before a strong earthquake in their model catalog. Carlson (1991), Huang et al. (1992), and Narkounskaia et al. (1992) considered a two-dimensional variant of the slider-block model.

Bak et al. (1987, 1988) suggested a simple cellular automaton-type ("sandpile") model represented by a lattice of threshold elements with random loading and a simple deterministic rule of stress release and nearest-neighbor redistribution. A sequence of consecutive breaks in the stress redistribution phase of the model was called an *avalanche*. The model is mathematically equivalent to a variant of the slider-block model in the limit of zero-mass blocks, and the avalanches can be interpreted as the earthquakes in the Burridge-Knopoff model. The sandpile model demonstrates an important property of *self-organized criticality*: from any initial state it evolves to a critical state characterized by a power law distribution of the avalanche sizes and two-point correlations. The applications of this model and its different variations and modifications can be found in Bak and Tang (1989), Ito and Matsuzaki (1990), Nakanishi (1991), Brown et al. (1991), Lomnitz-Adler et al. (1992), Vasconcelos et al. (1992), and Olami et al. (1992). See Ito (1992) for a review.

These models are concerned, in particular, with the power law distribution of earthquake sizes and, in general, with the chaotic character of a simple, homogeneous, and often deterministic, system. Different macroscopic effects due to changes in the local interaction rules, and phase transition phenomena according to variation of parameters were also investigated.

Although these models are rather abstract and oversimplified, some important features of seismicity can be understood in these models, and the influence of different types of interaction on the model catalog can be easily verified. It is important also as a possibility to establish analogies between the problems of predictability in solid Earth geophysics and other sciences.

HIERARCHICAL AND FRACTAL STRUCTURES

Models of crack nucleation based on the hierarchical block structure of the Earth's lithosphere were suggested in Allègre et al. (1982), Knopoff and Newman (1983), Smalley et al. (1985), Narkunskaya and Schnirman (1990), Molchanov et al. (1990), Newman and Gabrielov (1991), Gabrielov and Newman (1991), and Tumarkin and Shnirman (1992). All of these models explicitly introduce fractures of several scales and apply renormalization group methods to study interrelations between different scales. The condition for failure sometimes appears in these models as a critical phenomenon. In Smalley et al. (1985) and Newman and Gabrielov (1991), this approach explains the apparent low strength of fault zones—however, a critical point for failure does not emerge. Narkunskaya and Schnirman (1990) suggested a precursory pattern "upward bend of the frequency law" for major failures based on an analytic and numerical study of their model. This pattern has been later found in catalogs of seismicity for several regions.

King (1983) suggested that the kinematic incompatibility of the motion of lithospheric blocks was the source of fractal structure in the lithosphere in King (1983). Fractality is a general pattern of finite brittle strain in different materials [Turcotte, 1986 and 1990; Sornette et al., 1990; Sornette, 1991; King and Sammis, 1992].

INTERACTION OF TECTONIC FAULTS

There are a few models of seismicity where the interaction of tectonic faults is taken into account. One is the fluctuation model due to Rundle (1988) where earthquakes are treated as small thermodynamic fluctuations in the steady tectonic loading process in an elastic medium with embedded fault patches. Another is the block model [Gabrielov et al., 1990] where a seismically active region is modeled as a system of rigid blocks of arbitrary geometry separated by thin layers that represent fault zones. In Gabrielov et al. (1990), an algorithm known as *CN* for intermediate-term earthquake prediction is successfully applied to a model catalog. More recently, Yamashita and Knopoff (1992) suggested interaction in a system of faults as a mechanism for the activation-quiescence pattern.

DISCUSSION

The physical mechanism for earthquakes and other influence on subsequent events ("aftereffects") is still not fully understood. Adequate modeling of these aftereffects is important for the earthquake prediction problem, particularly because abnormal post-seismic activation is one of the intermediate-term premonitory patterns. Many of the models of seismicity in earthquake prediction studies associate premonitory patterns with the processes in the Earth's lithosphere on an intermediate time scale (i.e. between tectonic loading and elastic stress drop). These processes have been studied, mainly in laboratory experiments, and are not yet well understood for the solid Earth. A fundamental question is whether and what type of scaling may exist between laboratory samples and real tectonic faults. Most of the existing models of seismicity do not include the spatial distribution of earthquakes or are restricted to a single fault plane. At the same time, many of the premonitory seismicity patterns, as well as earthquake aftereffects, are observed far away from the fault where a major earthquake takes place. This means that the modeling of the interaction of the processes in different tectonic faults is important for understanding the sequence of events leading up to an earthquake.

Two important problems make it difficult to include real fault geometry in models of seismicity. First, to produce a long time model catalog with properties that stay unchanged in time, we need a stationary process in the model. Stationarity is an important ingredient of most earthquake prediction methods, providing the possibility to transfer the previously observed patterns to possible future events. At the same time, the underlying tectonics cannot be stationary, due to simple geometric considerations [McKenzie and Morgan, 1969; King, 1983]. Tectonic faults tend to grow in time. The kinematic instability of fault junctions leads to the creation and growth of complicated fractal structures—such as "morphostructural nodes" [Alekseevskaya et al., 1977]—around existing junctions and, eventually, to the emergence of new major faults. This non-stationarity appears as a fundamental challenge to the modeling of seismicity based on the actual fault's geometry.

Second, fault systems have a hierarchical fractal structure, and premonitory seismicity patterns are usually based on the properties of weak and moderate earthquakes that appear in the lower levels of this hierarchy. It is impossible in practical terms to handle several levels of this hierarchy in an explicit way. So, the problem that emerges is how to combine the available information on the geometry of principal faults with what is essentially statistical data on the fault system as a whole.

The challenge ahead is to develop an adequate language of description for the seismotectonic process. The language of continuum mechanics cannot describe the combination of continuous and discrete features of the seismotectonic process, namely its self-similar multi-scale spatial and temporal nature. However, the language of statistical physics and nonlinear science can describe complicated universal phenomena, but does not accommodate the specific geometry of individual tectonic faults. What is required, therefore, is a synthesis of these traditional and new approaches.

Acknowledgements. We wish to thank V.I. Keilis-Borok and D.L. Turcotte for many interesting discussions. A.M.G. wishes to thank the Department of Geological Sciences at Cornell University for its hospitality; his contribution was supported under NSF #EAR-91-04624.

REFERENCES

Alekseevskaya, M.A., A.M. Gabrielov, A.D. Gvishiani, I.M. Gelfand, and E.Ya. Rantsman, 1977. "Formal morphostructural zoning of mountain territories," *J. Geophys.*, **43**, 227–233.

Aki, K., 1979. "Characterization of barriers on an earthquake fault," *J. Geophys. Res.*, **84**, 6140–6148.

Aki, K., 1984. "Asperities, Barriers, Characteristic Earthquakes and Strong Motion Prediction," *J. Geophys. Res.*, **89**, 5867–5872.

Aki, K., 1987. "Magnitude-frequency relation for small earthquakes; a clue to the origin of f_{max} of large earthquakes," *J. Geophys. Res.*, **92**, 1349–1355.

Allègre, C.J., J.L. Le Mouel, and A. Provost, 1982. "Scaling rules in rock fracture and possible implications for earthquake prediction," *Nature*, **297**, 47–49.

Anderson, D.L., 1975. "Accelerated plate tectonics," *Science*, **187**, 1077–1079.

Anderson, O.L., and P.C. Grew, 1977. "Stress corrosion theory of crack propagation with applications to geophysics," *Rev. Geophys. Space Phys.*, **15**, 7–104.

Atkinson, B.K, 1984. "Subcritical crack growth in geological materials," *J. Geoph. Res.*, **89**, 4077–4114.

Bak, P. and C. Tang, 1989. "Earthquakes as a self-organized critical phenomenon," *J. Geophys. Res.*, **94**, 15635–15637.

Bak, P., C. Tang, and K. Wiesenfeld, 1987. "Self-organized criticality: an explanation of $1/f$ noise," *Phys. Rev. Lett.*, **59**, 381–384.

Bak, P., C. Tang, and K. Wiesenfeld, 1988. "Self-organized criticality," *Phys. Rev. A*, **38**, 364–374.

Bakun, W.H. and T.V. McEvilly, 1984. "Recurrence models and Parkfield, California, earthquakes," *J. Geophys. Res.*, **89**, 3051–3058.

Ben-Zion, Y., J.R. Rice, and R. Dmowska, 1993. "Interaction of the San Andreas fault creeping segment with adjacent Great Rupture zones and earthquake recurrence at Parkfield," *J. Geophys. Res.*, **98**, 2135-2144.

Bonin, J., Editor, 1991. "Earthquake predictions: state-of-the-art,"

Proc. of the Intnl. Conf., Strasbourg, France, 15-18 October, 1991. A.A. Balkema, Rotterdam.

Brown, S.R., C.H. Scholz, and J.B. Rundle, 1991. "A simplified spring-block model of earthquakes," *Geophys. Res. Lett.*, **18**, 217–218.

Burridge, R. and L. Knopoff 1967. "Model and theoretical seismicity," *Bull. Seismol. Soc. Amer.*, **57**, 341–371.

Carlson, J.M., 1991. "A two-dimensional model of a fault," *Phys. Rev. A*, **44**, 6226–6232.

Carlson, J.M. and J.S. Langer, 1989. "Mechanical model of an earthquake fault," *Phys. Rev. A*, **40**, 6470–6484.

Carlson, J.M., J.S. Langer, B.E. Shaw, and C. Tang, 1991. "Intrinsic properties of a Burridge-Knopoff model of an earthquake fault," *Phys. Rev. A*, **44**, 884–897.

Cao, T., and K. Aki, 1986. "Seismicity simulation with a rate- and state-dependent law," *PAGEOPH*, **124**, 487–514.

Das, S., and K. Aki, 1977. "Fault planes with barriers: a versatile earthquake model," *J. Geophys. Res.*, **82**, 5648–5670.

Das, S., and C.H. Scholz, 1981. "Theory of Time-Dependent Rupture in the Earth," *J. Geophys. Res.*, **86**, 6039–6051.

Dieterich, J.H., 1972. "Time-dependent friction in rocks," *J. Geophys. Res.*, **77**, 3690–3697.

Elsasser, W.M., 1969. "Convection and stress propagation in the upper mantle," in S.K. Runcorn (ed.), *The Application of Modern Physics to the Earth and Planetary Interiors* (New York: Wiley Interscience) 223-246.

Gabrielov, A.M., and V.I. Keilis-Borok, 1983. "Patterns of Stress Corrosion: Geometry of the Principal Stresses," *PAGEOPH*, **121**, 477–494.

Gabrielov, A.M., T.A. Levshina, and I.M. Rotwain, 1990. "Block model of earthquake sequences," *Phys. Earth Planet. Int.*, **61**, 18–28.

Gabrielov, A.M., and W.I. Newman, 1991. "Failure of Hierarchical Distributions of Fiber Bundles. II". submitted to *J. Appl. Probability*.

Gu, J., J.R. Rice, A.L. Ruina, and S.T. Tse, 1984. "Slip motion and stability of a single degree of freedom elastic system with rate and state dependent friction," *J. Mech. Phys. Solids*, **32**, 167–196.

Gutenberg, M, and C.F. Richter, 1944. "Frequency of Earthquakes in California," *Bull. Seismol. Soc. Amer.*, **34**, 185–188.

Huang, J., G. Narkounskaia, and D.L. Turcotte, 1992. A cellular-automata, slider-block model for earthquakes. 2. Demonstration of self-organized criticality for a 2D system. Geophys. J. International, **111**, 259-269.

Ishimoto, M., and K. Iida, 1939. Observations sur les Seismes Enregistres par le Microsismographe Construit Dernierement (1). *Bull. Earthquake Res., Inst.* Tokyo Univ., **17**, 443–478 (in Japanese).

Ito, K., 1992. "Towards a new view of earthquake phenomena," *PAGEOPH*, **138**, 531–548.

Ito, K., and M. Matsuzaki, 1990. "Earthquakes as self-organized critical phenomena," *J. Geophys. Res.*, **95**, 6853–6860.

Kagan, Y.Y., and L. Knopoff, 1980. "Spatial Distribution of Earthquakes: The Two-Point Correlation Function," *Geophys. J. Roy. astr. Soc.*, **62**, 303–320.

Kagan, Y.Y., 1991. "Fractal dimension of brittle fracture," *J. Nonlinear Sci.*, **1**, 1–16.

Kanamori, H. and G.S. Stewart, 1978. "Seismological aspects of the Guatemala earthquake of February 4, 1976," *J. Geophys. Res.*, **83**, 3427–3434.

Keilis-Borok, V.I., L. Knopoff, and I.M. Rotwain, 1980. "Bursts of aftershocks, long-term precursors of strong earthquakes," *Nature.*, **283**, 258–263.

Keilis-Borok, V.I. (ed.), 1990a. "Intermediate-Term Earthquake Prediction: Models, Algorithms, Worldwide Tests," *Phys. Earth Planet. Int.*, Special Issue, **61**, n.1-2.

Keilis-Borok, V.I., 1990b. "The lithosphere of the Earth as a nonlinear system with implications for earthquake prediction," *Rev. Geophys.*, **28**, 19–34.

Keilis-Borok, V.I. and I.M. Rotwain, 1990. "Diagnosis of time of increased probability of strong earthquakes in different regions of the world: algorithm CN," *Phys. Earth Planet. Int.*, **61**, 57–72.

Keilis-Borok, V.I. and V.G. Kossobokov, 1990. "Premonitory activation of earthquake flow: algorithm M8," *Phys. Earth Planet. Int.*, **61**, 73–83.

King, G.C.P., 1983. "The accomodation of large strains in the upper lithosphere of the Earth and other solids by self-similar fault systems: the geometrical origin of b-value," *PAGEOPH*, **121**, 761–815.

King, G.C.P., 1986. "Speculations on the geometry of the initiation and termination processes of earthquake rupture and its relation to morphology and geological structure," *PAGEOPH*, **124**, 567–585.

King, G. and J. Nabelek, 1983. "Role of fault bends in the initiation and termination of earthquake rupture," *Science*, **228**, 984–987.

King, G.C.P. and C.G. Sammis, 1992. "The mechanisms of finite brittle strain," *PAGEOPH*, **138**, 611–640.

Knopoff, L., and W.I. Newman, 1983. "Crack Fusion as a Model for Repetitive Seismicity," *PAGEOPH*, **121**, 495–510.

Lehner, F.K., V.C. Li, and J.R. Rice, 1981. "Stress diffusion along rupturing plate boundaries," *J. Geophys. Res.*, **86**, 6155–6169.

Li, V.C. and J.R. Rice, 1987. "Crustal deformation in great California earthquake cycles," *J. Geophys. Res.*, **92**, 11533–11551.

Lomnitz-Adler, J., L. Knopoff, and J. Martinez-Mekler, 1992. "Avalanches and epidemic models of fracturing in earthquakes," *Phys. Rev. A*, **45**, 2211–2221.

Lorenzetti, E., and T.E. Tullis, 1989. "Geodetic predictions of a Strike-Slip Fault Model: Implications for Intermediate- and Short-Term Earthquake Prediction," *J. Geophys. Res.*, **94**, 12343–12361.

Ma, Z., Z. Fu, Y. Zhang, C. Wang, G. Zhang, and D. Liu, 1990. *Earthquake prediction: Nine major earthquakes in China* (New York: Springer-Verlag).

Marone, C.J., C.H. Scholz, and R. Bilham, 1991. "On the Mechanics of Earthquake Afterslip," *J. Geophys. Res.*, **96**, 8441–8452.

McKenzie, D.P. and W.J. Morgan, 1969. "Evolution of triple junctions," *Nature*, **224**, 125–133.

Mikumo, T. and T. Miyatake, 1983. "Numerical modeling of space and time variations of seismic activity before major earthquakes," *Geophys. J. Roy. Astron. Soc.*, **74**, 559–583.

Mogi, K., 1968. "Migration of seismic activity," *Bull. Earthquake Res. Inst,*, Tokyo Univ., **46**, 53–74.

Mogi, K., 1969. "Some features of recent seismic activity in and near Japan (2): Activity before and after great earthquakes," *Bull. Earthquake Res. Inst,*, Tokyo Univ., **47**, 395–417.

Mogi, K., 1985. *Earthquake Prediction* (Tokyo: Academic Press).

Molchanov, S.A., V.F. Pisarenko, and A.Ya. Reznikova, 1990. "Multiscale models of failure and percolation," *Phys. Earth Planet. Int.*, **61**, n. 1–2, 36–43.

Nakanishi, H., 1991. "Statistical properties of the cellular-automaton model for earthquakes," *Phys. Rev. A*, **43**, 6613–6621.

Narkounskaia, G., J. Huang, and D.L. Turcotte, 1992. "Chaotic and self-organized critical behavior of a generalized slider-block model," *J. Stat. Phys.*, **67**, 1151–1183.

Narkunskaya, G.S., and M.G. Schnirman, 1990. "Hierarchical model of defect development and seismicity," *Phys. Earth Planet. Int.*, **61**, 29–35.

Newman, W.I. and L. Knopoff 1982. "Crack Fusion Dynamics: A Model for Large Earthquakes," *Geophys. Res. Lett.*, **9**, 735–738.

Newman, W.I. and L. Knopoff 1982. "A Model for Repetitive Cycles of Large Earthquakes," *Geophys. Res. Lett.*, **10**, 305–308.

Newman, W.I., and L. Knopoff, 1990. "Scale invariance in brittle fracture and the dynamics of crack fusion," *Int. J. Fracture*, **43**, 19–24.

Newman, W.I. and A.M. Gabrielov, 1991. "Failure of Hierarchical Distributions of Fiber Bundles. I," *Int. J. Fracture*, **50**, 1–14.

Nishenko, S.P., and R. Buland, 1987." A generic recurrence interval distribution for earthquake forecasting," *Bull. Seismol. Soc. Amer.*, **77**, 1382–1399.

Ogata, Y., 1992. "Detection of precursory relative quiescence before great earthquakes through a statistical model," *J. Geophys. Res.*, **97**, 19845–19871.

Olami, Z., H.J.S. Feder, and K. Christensen, 1992. "Self-organized criticality in a non-conservative cellular automaton modeling earthquakes," *Phys. Rev. Lett.*, **68**, 1244–1247.

Omori, F., 1895. "On the Aftershocks of Earthquakes," *Tokyo Imper. Univ.*, **7**, 111–200 (with Plates IV-XIX).

Reid, H.F., 1910. "Permanent displacements of the ground in The California Earthquake of April 18, 1906," Report of the State Earthquake Investigation Commission, vol. 2, pp.16–28, Carnegie Institution of Washington, Washington, D.C.

Rhebinder, P.A., and E.D. Shtchukin, 1972. "Surface phenomena affecting solids in the process of deformation and fracturing," *Uspekhi Fiz. Nauk*, **108**, 3–42 (in Russian).

Rice, J.R., 1980. "The mechanics of earthquake rupture," in A.M. Dziewonski and E. Boschi (eds.), *Physics of the Earth's Interior* (Amsterdam: North-Holland) 555–649.

Rice, J.R., 1993. "Spatio-temporal complexity of slip on a fault," *J. Geophys. Res.*, **98**, 9885–9907.

Rice, J.R., and J.-C. Gu, 1983. "Earthquake aftereffects and triggered seismic phenomena," *PAGEOPH*, **121**, 187–219.

Ruina, A., 1983. "Slip instability and state variables friction laws," *J. Geophys. Res.*, **88**, 10359–10370.

Rundle, J.B., 1988. "A physical model of earthquakes: 2. Applications to Southern California," *J. Geophys. Res.*, **93**, 6255–6274.

Scholz, C.H., 1990. *The mechanics of earthquakes and faulting*, Cambridge University Press, Cambridge.

Schwartz, D.P. and K.J. Coppersmith, 1984. "Fault behavior and characteristic earthquakes: Examples from the Wasatch and San Andreas faults," *J. Geophys. Res.*, **89**, 5681–5698.

Schwartz, D.P., K.J. Coppersmith, F.H. Swan III, P. Sommerville, and W.U. Savage, 1981. "Characteristic earthquake on intraplate normal faults," *Earthquake notes*, **51**, 71.

Shaw, B.E., J.M. Carlson, and J.S. Langer, 1992. "Patterns of seismic activity preceding large earthquakes," *J. Geophys. Res.*, **97**, 479.

Smalley, R.F., D.L. Turcotte, and S.A. Solla, 1985. "A renormalization group approach to the stick-slip behavior of faults," *J. Geophys. Res.*, **90**, 1894–1900.

Sornette, D., 1992. "Self-organized criticality in plate tectonics," in T. Riste and D. Sherrington (eds.), *Spontaneous Formation of Space-Time Structures and Criticality*, NATO Advanced Studies Institute Series C: Mathematical and Physical Sciences–Vol. 349, (Dordrecht: Kluwer Academic Publishers) 57–106.

Sornette, A., P. Davy, and D. Sornette, 1990. "Growth of fractal fault patterns," *Phys. Rev. Lett.*, **65**, 2266–2269.

Sornette, D., C. Vanneste, L. Knopoff, 1991. "Statistical model of earthquake foreshocks," *Phys. Rev. A*, **45**, 8351–8357.

Takayasu, H., 1990. *Fractals in the physical sciences* (Manchester: Manchester University Press).

Takayasu, H., and M. Matsuzaki, 1988. *Phys. Lett. A*, **131**, 244-247.

Thatcher, W., 1990. "Order and diversity in the models of circum-Pacific earthquake recurrence," *J. Geophys. Res.*, **95**, 2609–2623.

Tse, S. and J.R. Rice, 1986. "Crustal earthquake instability in relation to the depth variation of frictional slip properties," *J. Geophys. Res.*, **91**, 9452–9472.

Tumarkin, A.G. and M.G. Shnirman, 1992. "Critical Effects in Hierarchical Media," *Computational Seismology*, **25**, 63-71 (in Russian).

Turcotte, D.L., 1986. "Fractals and Fragmentation," *J. Geophys. Res.*, **91**, 1921–1926.

Turcotte, D.L., 1992. *Fractals and chaos in geology and geophysics* (Cambridge: Cambridge University Press).

Vasconcelos, G.L., M.D. Vieira, and S.R. Nagel, 1992. "Phase-transitions in a spring block model of earthquakes," *Physica A*, **191**, 69–74.

Wyss, M. and R.E. Habermann, 1988. "Precursory seismic quiescence," *PAGEOPH*, **126**, 319–332.

Yamashita, T. and L. Knopoff, 1987. "Models of aftershock occurrence," *Geophys. J. Roy. astr. Soc.*, **91**, 13–26.

Yamashita, T. and L. Knopoff, 1989. "A model of foreshock occurrence," *Geophys. J. Roy. astr. Soc.*, **96**, 389-399.

Yamashita, T. and L. Knopoff, 1992. "Model for intermediate-term precursory clustering of earthquakes," *J. Geoph. Res.*, **97**, 19873–19879.

Andrei Gabrielov, Department of Mathematics, University of Toronto, Toronto, Canada M5S 1A1

William I. Newman, Department of Earth and Space Sciences, University of California, 4640 Geology Building, Los Angeles, CA 90024.

Nonlinear Dynamics of the Transition from Stable Sliding to Cyclic Stick-Slip in Rock

YAOJUN GU AND TENG-FONG WONG

Department of Earth and Space Sciences, State University of New York at Stony Brook

ABSTRACT: The nonlinear dynamical behavior in the transition from stable sliding to cyclic stick-slip in three rock-gouge systems (Westerly granite with ultrafine quartz gouge, sawcut Tennessee sandstone, and Tennessee sandstone with natural halite gouge) were investigated in a conventional triaxial configuration at confining pressures from 30 MPa to 110 MPa. Induced by cumulative slip or load point velocity perturbation, the transition from stable sliding to cyclic stick-slip might involve several distinct sliding modes, including chaotic oscillations, period doubling and self-sustained oscillations. The accumulation of slip may destabilize or stabilize the frictional sliding behavior, whereas an increase of load point velocity tends to stabilize the oscillational behavior. The experimental observations were interpreted using a spring-slider model with single degree of freedom. To reproduce the full range of sliding modes and their observed dependence on velocity perturbation, it was necessary to incorporate two state variables and high-speed cutoff of velocity dependence in the Dieterich-Ruina rate- and state-dependent friction law. Analytic expression for the onset of Hopf bifurcation in such a system was derived. Stability boundaries dependent on stiffness, load point velocity, and friction constitutive parameters were established based on limit cycle oscillation modes observed in numerical simulations. Critical condition for the triggering of dynamic instability by sudden velocity perturbation was analyzed. Seismological implications of our experimental observations and spring-slider simulations were also discussed.

INTRODUCTION

Since *Brace and Byerlee* [1966] proposed that stick-slip instability between sliding rock surfaces in the laboratory provides an analogue for crustal earthquakes, the study of frictional sliding behavior has been extensively pursued in the laboratory. Experimental rock friction studies have provided important physical insights on earthquake source mechanics (see reviews by *Byerlee* [1978], *Rice* [1983], and *Scholz* [1990]). In the laboratory, there are two primary modes of frictional sliding. In the *stable sliding* mode, the two surfaces slip steadily at a relative velocity equal to the load point velocity, and this behavior is probably an analogue of fault creep [*Byerlee and Summers*, 1975]. In the *stick-slip* mode, the frictional surfaces suddenly slip, lock and then slip in a cyclic manner, and this behavior is an analogue of the earthquake cycle along a preexisting crustal fault [*Brace and Byerlee*, 1966].

In the laboratory, whether frictional sliding occurs in a stable manner or by cyclic stick-slip depends in a complex way on normal stress, temperature, mineralogy, amount of gouge, presence of fluid, surface geometry, and possibly other factors. In general, the transition from stable sliding to cyclic stick-slip is promoted by high normal stresses, by low temperatures, by the presence of strong, brittle minerals, and by lower surface roughness [*Stesky*, 1978]. The sliding mode is also sensitive to velocity perturbation and time of stationery contact [*Dieterich*, 1978; *Ruina*, 1983; *Tullis and Weeks*, 1986]. Under conditions favorable for the occurrence of stick-slip, cumulative slip is usually observed to have a destabilization effect in that a certain amount of stable sliding usually precedes stick-slip [*Byerlee and Summers*,1975]. However, if the cyclic stick-slip results in significant wear along the gouge layer, then cumulative slip may also stabilize the sliding behavior resulting in a reverse transition from cyclic stick-slip to stable sliding [*Wong and Zhao*, 1990; *Wong et al.*, 1992].

Systematic investigations of the quasi-static sliding behavior of frictional surfaces have revealed that rock friction depends on slip rate and has memory. *Dieterich*

[1979] and *Ruina* [1983] formulated a rate- and state-dependent friction law capable of reproducing the key attributes of the quasi-static sliding behavior. Furthermore, this friction law can provide a self-consistent model for the cyclical nature of stick-slip and characteristic earthquakes [*Tse and Rice*, 1986]. Recent theoretical analyses [*Ruina*, 1983; *Gu et al.*, 1984] have shown that in a spring-slider system obeying the Dieterich-Ruina rate- and state-dependent friction law, the transition from stable sliding to cyclic stick-slip should go through a route involving chaotic oscillations, period doubling and self-sustained, periodic oscillations. Such nonlinear dynamical phenomena have important implications on the generation of creep waves and on the periodicity of characteristic earthquakes along seismogenic zones [*Scholz*, 1990]. The details of the transitional behavior are of interest in nonlinear dynamics and they provide important constraints on the friction constitutive relation [*Wong et al.*, 1992].

The experimental study of frictional sliding behavior has historically focused on the two end members, with little attention on the details of the transition from stable sliding to cyclic stick-slip. In this paper, we will first review our experimental observations on the nonlinear dynamical behavior in the transitional regime. The frictional sliding behavior of three different rock-gouge sytems (Westerly granite with ultrafine quartz gouge, sawcut Tennessee sandstone, and Tennessee sandstone with halite gouge) were investigated, focusing on the transition induced by cumulative slip and by load point velocity perturbation.

We will compare the experimental observations with theoretical predictions, specifically on the complexity in dynamical behavior which results from the nonlinearity of the Dieterich-Ruina friction constitutive relation. An appropriate model for the common configurations in rock friction experiments [*Jaeger and Cook*, 1971; *Paterson*, 1978; *Scholz*, 1990] is the spring-loaded block-slider with a single degree of freedom (Figure 1). Such a vibrating system is of importance in the modeling of many mechanical, electrical and acoustic systems [*Rayleigh*, 1896; *Stoker*, 1950]. In earthquake mechanics, this model has recently been used to analyze triggered seismic phenomena [*Rice and Gu*, 1983], earthquake cycle [*Rice and Tse*, 1986], earthquake nucleation [*Dieterich*, 1986], and earthquake afterslip [*Marone et al.*, 1991].

In nonlinear dynamics, stable sliding and cyclic stick-slip correspond respectively to a stable point attractor and a periodic orbit in the phase portrait, and the transition from stable sliding to cyclic stick-slip is an example of the "Hopf bifurcation" [*Thompson and Stewart*, 1986]. Analytic results for the onset of Hopf bifurcation in a spring-slider system obeying the Dieterich-Ruina friction

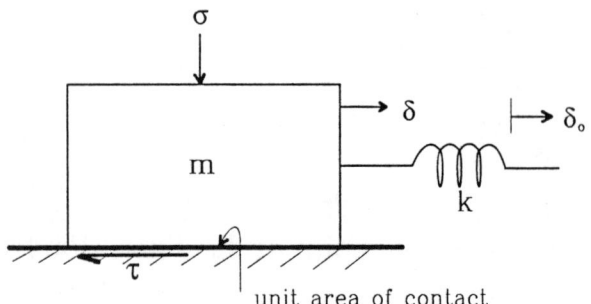

Fig. 1. Single degree of freedom spring-slider system. The slider of unit area and mass m is subjected to a constant normal stress σ, friction stress τ, and stressing of the spring with a stiffness of k. Displacements of the slider and the load point are δ and δ_o respectively.

law can be derived. We summarize below these analytical solutions, as well as the nonlinear oscillational behavior associated with the bifurcation which can only be investigated by numerical simulations. In particular, the variation in sliding behavior due to the incorporation of multiple state variables and high-speed cutoff of the velocity dependence in the Dieterich-Ruina friction law will be considered.

While a systematic comparison of the experimental observations with theoretical analyses provides important physical insights on the friction constitutive relation and the nonlinear dynamical behavior of a seismogenic system, we do not suggest that our highly simplified model corresponds exactly to an actual fault. The dynamics of complex systems and the effects of mechanical and geometric heterogeneities on earthquake faulting are two topics being intensely investigated using discrete [e.g. *Cao and Aki*, 1986; *Bak and Tang*, 1989; *Sornette and Sornette*, 1989; *Nakanishi*, 1990; *Carlson and Langar*, 1989; *Brown et al.*, 1991] and continuum [*Horowitz and Ruina*, 1989; *Rice*, 1991; *Madariaga et al.*, 1991] models. We will discuss the seismological implication of our experimental data in light of these recent results.

TRANSITION FROM STABLE SLIDING TO CYCLIC STICK-SLIP: EXPERIMENTAL OBSERVATIONS

Experimental rock friction research of the past three decades have been reviewed by *Byerlee* [1978], *Paterson* [1978], *Tullis* [1986], and *Scholz* [1990]. Three principal experimental configurations (direct shear, rotary shear and triaxial compression) are commonly used. In our study, the frictional sliding experiments were performed in the conventional triaxial configuration with three types of improvement incorporated into the test system. First, an internal load cell was included as an integral part of a newly designed endplug. The internal load cell was placed in direct contact with the sample assembly inside the pressure vessel, and it significantly enhances the resolution of the stress variations. Second, the displacement rate of the loading ram was servo-controlled and the function generator in the servo-control feedback loop was upgraded to 16-bit. This new feature allows us to readily change the reference displacement rate at relatively small increments. Third, three data acquisition systems of different capabilities were set up to monitor oscillational behavior with vastly different characteristic times. The first system acquires the load cell, pressure transducer, and displacement transducer signals at a relatively low rate (of several samples per second) through a 14-bit A/D convertor. The second system takes the signal from the internal load cell at a sampling rate up to 100 sample/s. The third system connects the load cell and displacement transducer to a digital storage oscilloscope for studying the details of selected dynamic instability processes at a sampling rate of 10 ns/word.

The cylindrical country rock samples were of diameter 25.4 mm and nominal length 63.5 mm containing a sawcut at 30° to the symmetry axis. The sliding surfaces were ground with a 100-grit wheel. The Westerly granite samples were cored from the same block used by *Wong et al.* [1992], and the Tennessee sandstone samples were from the same block used by *Scott and Nielsen* [1991]. The ultrafine quartz powder (median particle size 6.4 µm) was sandwiched between sawcut surfaces of Westerly granite to form a simulated gouge layer of nominal thickness 0.3 mm. The halite (non-iodized cooking salt of average grain size 0.25 mm) was sandwiched between sawcut surfaces of Tennessee sandstone to form a simulated gouge layer of nominal thickness 0.5 mm. Experiments were also conducted on sawcut Tennessee sandstone samples without any simulated gouge. Additional details on experimental procedure were provided by *Wong et al.* [1992] and *Gu and Wong* [1992].

Two types of experiments were conducted. In the first type, we focused on the effect of cumulative slip and the axial displacement rate was maintained constant at 1 µm/s. In the second type, the load point velocity was continually perturbed after frictional sliding had commenced. The *load point velocity* is defined as the slip displacement rate resolved along the sawcut surface (which is higher than the reference axial displacement rate by a factor of 1.15).

Effect of cumulative slip on the frictional sliding behavior: destabilization vs stabilization

In frictional sliding experiments, it is often observed that stable sliding precedes cyclic stick-slip [e.g. *Byerlee*, 1967; *Dieterich*, 1972; *Scholz et al.*, 1972], and this observation suggested to *Scholz et al.* [1972] that fault creep could provide a premonitory signal to an impending earthquake. The effect of normal stress, load point velocity and rock type on the destabilization distance (i.e. distance of stable sliding preceding the onset of cyclic stick-slip) have been investigated [e.g. *Byerlee and Summers*, 1975]. Measurements of the friction constitutive parameters in quasi-static experiments by *Dieterich* [1981], *Blanpied* [1989] and *Biegel et al.* [1989] also imply that cumulative slip tends to destabilize the sliding behaivor.

Our study has shown that the sliding mode does not evolve from stable sliding to cyclic stick-slip through a sharp transition. Usually a number of quasi-static oscillations would occur in the transitional regime. Figure 2 shows details of the supercritical oscillations (in Westerly granite samples sliding on ultrafine quartz gouge at a fixed confining pressure of 50 MPa and load point velocity of 1.15 µm/s) which were not resolvable until we made the improvements in our system discussed above. Stable sliding first evolved to *self-sustained oscillations* of increasing amplitude which were subaudible and periodic. Then a *period doubling* stage (with sequential pairs of large and small stress drop oscillations) occurred before the onset of cyclic stick-slip with audible acoustic emissions. Self-sustained oscillations were observed in all

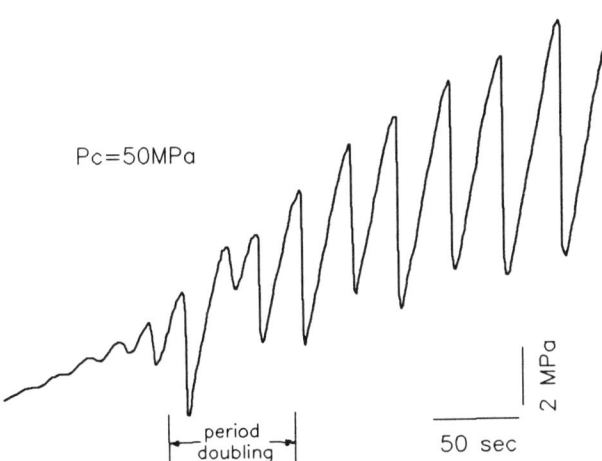

Fig. 2. Supercritical oscillations recorded by the internal load cell in experiment GU22 during the transition from stable sliding to cyclic stick-slip in Westerly granite sandwiched with ultrafine quartz gouge. Confining pressure was 50 MPa and load point velocity was 1.15 µm/s. Period doubling was observed in the destabilization phase.

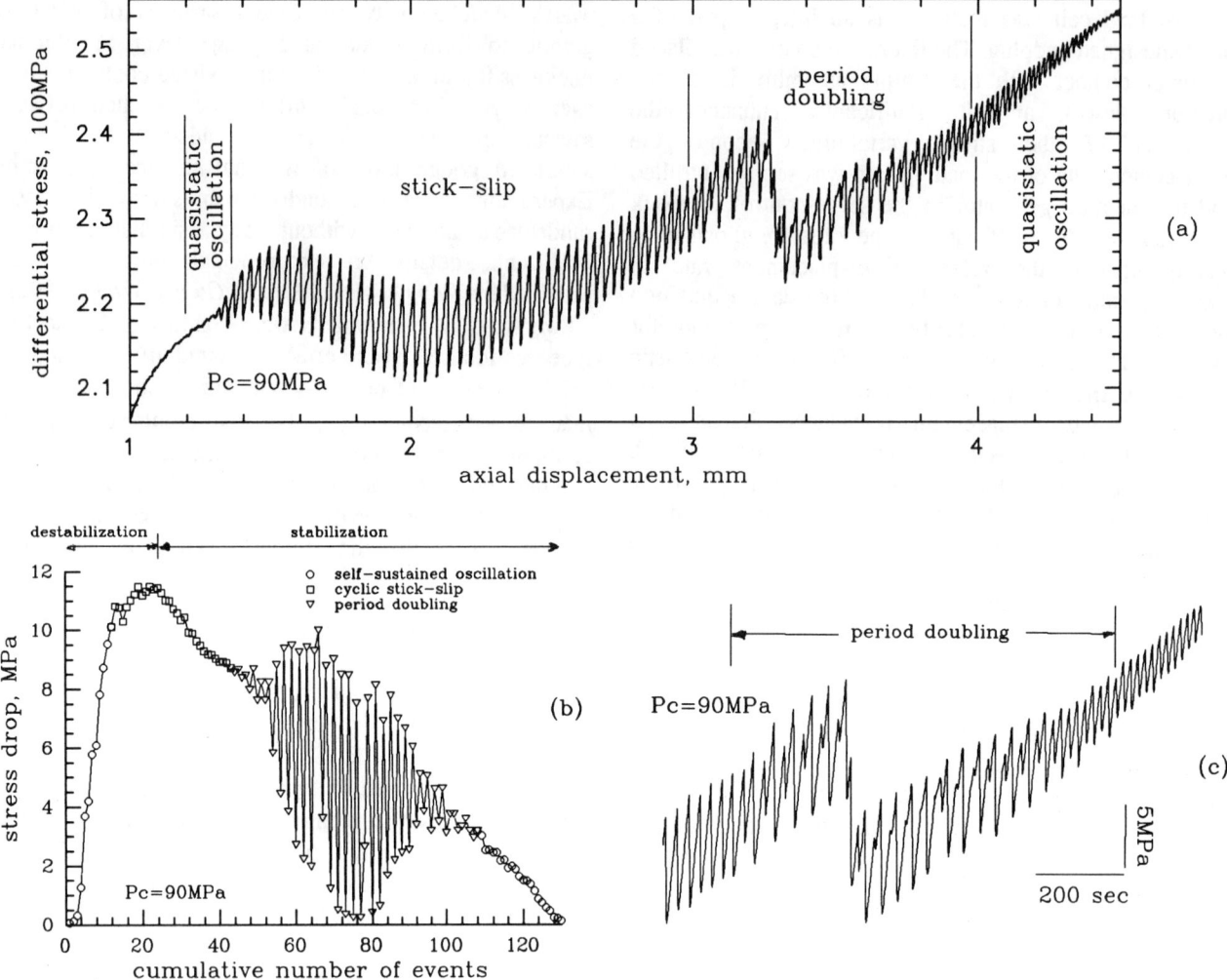

Fig. 3. (a) Complete record of the differential stress versus axial displacement in an experiment on Westerly granite sandwiched with ultrafine quartz gouge at confining pressure 90 MPa and axial displacement rate of 1 μm/s. (b) Differential stress drop amplitude as a function of cumulative number of events. The three sliding modes of cyclic stick-slip, period doubling, and self-sustained oscillation are highlighted. (c) Stress drops recorded by the internal load cell for axial displacement from 2.8 mm to 4.2 mm. The period doubling behavior is highlighted.

of our experiments on the ultrafine quartz gouge at confining pressures ranging from 50 MPa to 110 MPa [*Wong et al.* 1992], but period doubling was rarely observed at pressures above 50 MPa. In sawcut Tennessee sandstone samples, period doubling was more commonly observed [*Gu and Wong,* 1992].

Stress drop amplitudes of the cyclic stick-slip events were not constant. Typically, the stress drop would increase with cumulative slip until it reached a maximum value, and then decrease monotonically until the sliding mode evolved from cyclic stick-slip to period doubling (Figure 3a). The stick-slip event with the maximum stress drop therefore marked the termination of the *destabilization* stage as well as the onset of the *stabilization* stage. The period doubling evolved to self-sustained oscillations of decreasing amplitudes, which ultimately led to stable sliding. The scenario outlined here for the transition from stable sliding to cyclic stick-slip and then back to stable sliding is highlighted in a plot of the stress drop as a function of the cumulative number of events (Figure 3b) . This experiment was conducted on Westerly granite with ultrafine quartz gouge at a confining pressure 90 MPa. Period doubling was observed in the stabilization stage (Figure 3c) corresponding to the bifurcations in stress drop shown in Figure 3b, but it was absent in the destabilization stage.

In the stabilization stage, increasing normal stress tends to significantly decrease the amount of cumulative slip required for stabilization. *Wong et al.* [1992] attributed this to the enhancement of the wear rate at elevated normal stresses. The stabilization phenomenon has not been commonly observed in previous experiments on other rock-gouge systems. The relatively high hardness of quartz and the fine particle size of our simulated gouge might have contributed to shortening the distance over which significantly wear and stabilization occurred in our experiments. We have recently extended the observations in other materials. In sawcut Tennessee sandstone samples, we observed overall destabilization and stabilization trends qualitatively similar to the scenario outlined above, but the destabilization and cyclic stick-slip processes occurred over slip distances somewhat longer than those for Westerly granite with ultrafine qaurtz gouge. This may be related to differences between the two systems with regard to the rates by which wear and Riedel shear develop in the gouge layer [*Gu and Wong*, 1992].

The stabilization of faulting by cumulative slip is possibly related to recent seismological observations by *Wesnousky* [1990]. For 5 major strike-slip fault zones in southern California, he attempted to isolate the effect of cumulative slip on earthquake "productivity" by normalizing the number of earthquakes with respect to the fault length and the rate of slip accumulation. *Wesnousky* [1990] concluded that the seismic "productivity" decreases with increasing cumulative slip, and he suggested that the stabilization of faulting by cumulative slip may be related to his previous observation [*Wesnousky*, 1988] that the accumulation of geologic offset tends to reduce the structural complexity along a fault zone.

It should be noted that isolated observations of period doubling bifurcations and self-sustained oscillations were previously documented. Using a direct shear apparatus, *Scholz et al.* [1972] reported what was probably the first observation of self-sustained oscillations in sawcut granite surfaces without simulated gouge. Using a rotary shear apparatus, *Weeks and Tullis* [1985] observed self-sustained oscillations in dolomite. Using a direct shear apparatus, *Ruina* [1980] observed period doubling in quartzite when he perturbed the apparent stiffness of his servo-controlled test system.

Nonlinear dynamical behavior in response to velocity perturbation

Two aspects of the velocity dependence of frictional sliding behavior were extensively studied. The first is on the evolution of friction stress with slip when the velocity is suddenly perturbed during stable sliding. Such canonical experiments are crucial for the determination of constitutive parameters in the Dieterich-Ruina friction law [e.g. *Dieterich*, 1979; *Ruina*, 1983; *Tullis and Weeks*, 1986]. The second is on the variation in stress drop of cyclic stick-slip with respect ot the loading velocity. *Cao and Aki* [1986] suggested that this variation might provide an explanation of the seismological observation that stress drop increases with recurrance time [*Kanamori and Allen*, 1986].

If a frictional surface is in the stable sliding mode and the velocity is suddenly perturbed to a new value, and if the velocity perturbation is sufficiently small such that the sliding mode remains to be stable sliding, then the friction stress would first undergo an instantaneous change which is followed by an evolution change (Figure 4). In the rate- and state-dependent friction law formulated by *Dieterich* [1979] and *Ruina* [1983], three sets of constitutive parameters were introduced to respectively characterize the magnitude of the instantaneous and evolution changes in friction, and the characteristic slip distances over which the evolution changes occur. In addition, a set of internal variables are introduced to characterize phenomenologically the "state" of the frictional surface. For more than a decade, the type of quasi-static sliding experiments outlined in Figure 4 has been actively pursued [see review by *Tullis*, 1986]. Recent work has extended the scope of investigation to include a wide range of materials (e.g. dolomite, *Weeks and Tullis*, 1985; serpentine, *Reinen et al.*, 1991), hydrothermal conditions [*Blanpied et al.*, 1991; *Chester and Higgs*, 1992], and normal stress perturbations [*Linker and Dieterich*, 1992]. The overall framework of the Dieterich-Ruina friction law

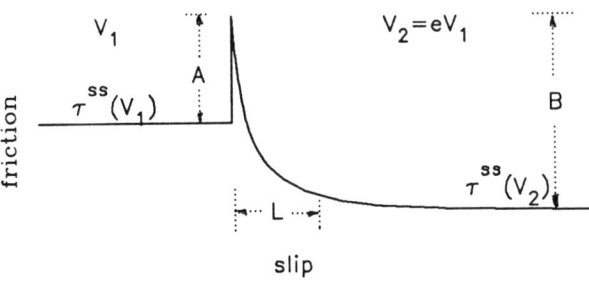

Fig. 4. Schematic diagram of the frictional behavior in response to a perturbation in sliding velocity in quasi-static experiments. For an e-fold perturbation in sliding velocity, there is an instantaneous increase in frictional stress with an amplitude of A and evolution decay with an amplitude of B over a characteristic length of L.

can be used to adequately reproduce the experimental data. The mathematical formulation and numerical values of the constitutive parameters will be elaborated in our latter discussion on theoretical analysis.

During cyclic stick-slip, extensive studies on a wide range of materials [e.g. *Rabinowicz*, 1958; *Ohnaka*, 1973; *Engelder et al.*, 1975; *Teufel and Logan*, 1978; *Shimamoto and Logan*, 1986] have shown that the stress drop amplitude decreases with increasing load point velocity. Variations in sliding behavior induced by velocity perturbations are coupled to the effect of cumulative slip which, as we emphasized above, is destabilizing in the first stage and stabilizing in the second stage in most of our experiments. Therefore, a quantitative investigation into the velocity perturbation effect should be restricted to relative small slip distances over which the destabilization or stabilization effect of cumulative slip is comparatively small. Recently, *Wong and Zhao* [1990] conducted a systematic study of this effect in Westerly granite with ultrafine quartz gouge. At a fixed normal stress, the relative change in friction coefficient in a stick-slip event was observed to vary linearly with the logarithm of the load point velocity [*Gu and Wong*, 1991].

Other than influencing the frictional strength during stable sliding and the stress drop of cyclic stick-slip, velocity perturbations can also exert significant influence on the sliding mode in the transition regime between stable sliding and cyclic stick-slip. Figure 5a shows our observations in the stabilization stage in Westerly granite with ultrafine quartz gouge. The sliding mode changed

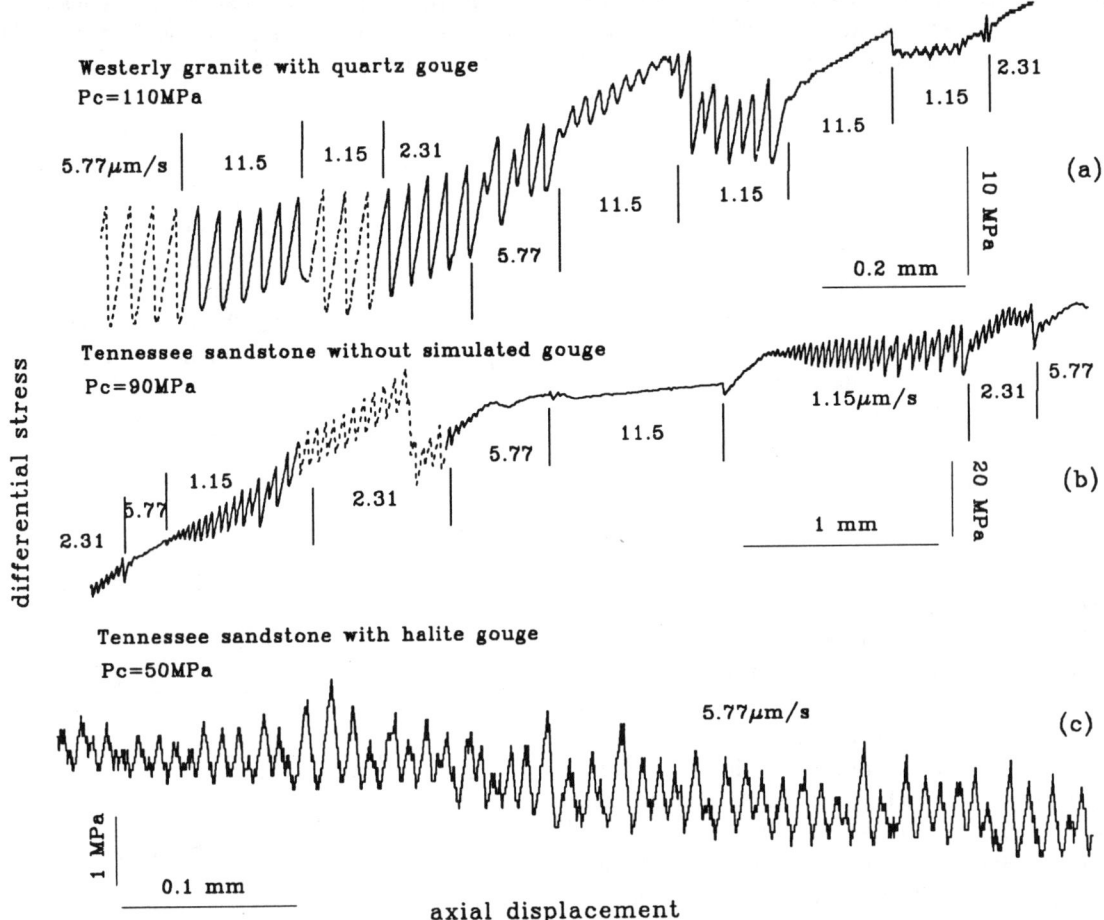

Fig. 5. (a) Transformation of frictional sliding mode in response to perturbations in load point velocity in Westerly granite with ultrafine quartz gouge at confining pressure 90 MPa. (b) Transformation of frictional sliding mode in response to perturbations in load point velocity in Tennessee sandstone without any simulated gouge at confining pressure 90 MPa. (c) Details of apparently chaotic behavior in Tennessee sandstone with halite gouge at confining pressure 50 MPa. The load point velocity (V_o) are indicated by the numbers along the curves. Cyclic stick-slip events with audible acoustic emissions are shown by the dashed curves, whereas stable sliding and quasi-static oscillations (including chaotic oscillations, period doubling and self-sustained, periodic oscillations) are shown by the solid curves.

from cyclic stick-slip to self-sustained oscillations as the load point velocity was perturbed from 5.77 µm/s to 11.5 µm/s, and the reverse change occurred in response to a perturbation from 11.5 µm/s to 1.15 µm/s. As the load point velocity increased from 2.31 µm/s to 5.77 µm/s and then to 11.5 µm/s, the sliding mode transformed from cyclic stick-slip to period doubling to self-sustained oscillation. In other words, increasing the load point velocity had a tendency to stabilize the frictional sliding behavior. The overall stabilization induced by cumulative slip can also be noted by comparing the three groups of data for 11.5 µm/s at different stages of slip.

Figure 5b shows data obtained towards the very end of the destabilization stage in sawcut Tennessee sandstone. Again the effect of increasing load point velocity was to stabilize the sliding behavior, and the transition from stable sliding to cyclic stick-slip involved period doubling and self-sustained oscillations. We also conducted experiments on Tennessee sandstone samples sandwiched with natural halite gouge. Previous studies by *Shimamoto and Logan* [1986] have shown that this system exhibits an usually broad spectrum of frictional sliding behavior at room temperature, possibly due to the complex interplay of cataclastic and crystal plasticity mechanisms which are sensitive to normal stress and strain rate [*Chester and Logan*, 1990]. Unlike the other two systems (Westerly granite with ultrafine quartz gouge and sawcut Tennessee sandstone), the sliding behavior of halite gouge is very sensitive to the loading history and the load point velocity. Therefore it is difficult to generalize our observations concerning the effects of cumulative slip on the sliding mode. As for load point velocity, the effect is similar to the other two systems in that the general trend was for a positive perturbation to stabilize the sliding behavior. One outstanding feature of this system is that it exhibited oscillations with stress drops which were apparently chaotic. Figure 5c highlights the details of such chaotic behavior.

STABILITY BEHAVIOR AND THE ONSET OF HOPF BIFURCATION

Dieterich-Ruina friction constitutive relation: multiple state variables and high-speed cutoff of velocity dependence

As we discuss above, an attractive feature of the Dieterich-Ruina type of rate- and state-dependent friction constitutive relation is that it can be used to model the cyclical behavior of stick-slip and the various stages of the earthquake cycle. The constitutive parameters of a given frictional system are to be determined from quasi-static sliding experiments. The canonical experiment focuses on the evolution of frictional strength in response to a sudden perturbation in slip velocity. If frictional sliding is stable at a velocity V and if there is a step increase in the velocity (say, by an amount $eV = 2.71828..V$), then the usual response is for the friction stress to increase instantaneously by a magnitude A and then decrease by an evolution change of magnitude B to a residual value over a characteristic slip distance L (Figure 4). If the magnitude of B is greater (or less) than A, then the frictional sliding behavior is called velocity weakening (or strengthing). The parameters A and B seem to be a function of the roughness and gouge particle size distribution. The "state" of the frictional surface is characterized by an internal variable θ, and it is commonly assumed that experimental measurements of L have to be scaled up by several orders of magnitude to apply to crustal faults [*Rice*, 1983; *Scholz*, 1990].

Systematic measurements [e.g. *Ruina*, 1980; *Weeks and Tullis*, 1985; *Tullis and Weeks*, 1986; *Cox*, 1990] have indicated that the evolution change often occurs over a multiplicity of characteristic slip distances (L_i, with $i = 1, 2, ... N$), each of which corresponding to a state variable θ_i and an evolution change of magnitude B_i. The friction constitutive relation is then given by the following general form:

$$\tau = \tau_* + A\ln(\frac{V}{V_*}) + \sum_{i=1}^{N} \theta_i, \quad (1a)$$

$$\frac{d\theta_i}{dt} = -\frac{V}{L_i}\left[\theta_i + B_i \ln(\frac{V}{V_*})\right]. \quad (1b)$$

where τ is the friction stress, and τ_* is its steady state value at a reference velocity V_*. It is expected that increasing the number of state variables would increase the complexity of the frictional sliding behavior. Most published data can be fitted quite well with at most 2 state variables. The general form (1) of the Dieterich-Ruina friction relation assumes that the rate dependence is the same over all velocities. However, experiments conducted over a wide range of velocities suggest that there exists a "high-speed cutoff" in velocity dependence. This effect has been observed in metals [e.g. *Richardson and Nolle*, 1976], rocks [*Dieterich*, 1978] and ice [*Jones et al.*, 1991]. *Okubo and Dieterich* [1986] have also emphasized the necessity of including this effect in the interpretation of their dynamic instability data. Motivated by these experimental observations, we use a rate- and state-dependent friction law with two (velocity weakening) state variables and with high-speed cutoff of the velocity dependence. This requires synthesizing the formulations of *Ruina* [1983] and *Rice and Tse* [1986] to arrive at the following specific form:

$$\tau = \tau_* + \theta_1 + \theta_2 + A \ln(\frac{V}{V_*}), \qquad (2a)$$

$$\frac{d\theta_1}{dt} = -\frac{V}{L_1}\left[\theta_1 - (B_1 - \frac{A}{2})\ln(\frac{V_*}{V} + e^{-n}) + \frac{A}{2}\ln(\frac{V}{V_*})\right], \qquad (2b)$$

$$\frac{d\theta_2}{dt} = -\frac{V}{L_2}\left[\theta_2 - (B_2 - \frac{A}{2})\ln(\frac{V_*}{V} + e^{-n}) + \frac{A}{2}\ln(\frac{V}{V_*})\right]. \qquad (2c)$$

As was discussed by *Rice and Tse* [1986], the above constitutive relation implies that the velocity dependence is cut off at speeds higher than $V_* e^n$, and therefore if n becomes infinitely large, the friction law reduces to the formulation with two state variables but without any high-speed cutoff. We believe that the above friction constitutive relation captures the key features of most quasi-static experiments. Nevertheless, there are certain complexities reflected in recent experiments which are not included here. These include a possible change of sign of the velocity dependence above a threshold velocity [*Blanpied et al.*, 1987; *Shimamoto and Logan*, 1986] and nonlinear dependence of the friction stress on the normal stress [*Linker and Dieterich*, 1992]. Incorporating these features into the formulation would complicate the mathematical treatment to an extent which is not warranted at our current stage of understanding of rock friction.

Analytic results for the onset of Hopf bifurcation

We consider the dynamics of a spring-slider system with a single degree of freedom (Figure 1). The slider has unit area and mass m, and the stiffness (spring constant) is k. The spring applies a force on the slider which is proportional to the displacement of the slider (δ) relative to that of the load point (δ_o), and the slider moves when this spring force overcomes the frictional resistance τ. The equation of motion of this system is therefore

$$m\frac{dV}{dt} = k(\delta_o - \delta) - \tau \qquad (3)$$

where time is denoted by t, and $V = d\delta/dt$ and $a = dV/dt$ are respectively the velocity and acceleration of the slider. Substituting equation (3) into (2), we obtain the governing equations for the dynamics of this spring-slider systems. The mathematical analysis is simplified by normalizations suggested by *Gu et al.* [1984] and the details are provided in the Appendix. Using *Rice's* [1983] formulation, only one of the two state variables needs to be included explicitly in the governing equations to arrive at a system of 4 ordinary differential equations for the velocity, acceleration, friction stress, and the first state variable (equation (i) of the Appendix).

The choice of the parameters for the spring-slider system and for the friction constitutive equations have to be guided by experimental rock friction data and physical characteristics of earthquake faulting. In the context of the Dieterich-Ruina friction law with one state variable and without high-speed cutoff, *Gu and Wong* [1991] considered a broad range of parameter values appropriate for crustal faulting and rock mechanics experiments. In this study, we add a second state varaible and also high-velocity cutoff of velocity dependence, rendering the mathematics to be more complicated. We have to consider four normalized friction constitutive parameters ($\rho = L_1/L_2$, $\beta_1 = B_1/A$, $\beta_2 = B_2/A$, and n) instead of only one normalized parameter ($\lambda = (B-A)/A$) considered by *Gu and Wong* [1991]. We have conducted simulations using different values for the normalized friction constitutive parameters. However, since in this paper we are primarily interested in the effects of the introduction of a second state variable and of the high-speed cutoff, we can highlight the major conclusions with reference to our results for only one set of values of ρ, β_1, and β_2. To simplify the presentation and to establish contact with the previous quasi-static study of *Gu et al.* [1984], we will focus on our results for the choice of parameters ($\rho = 0.048$, $\beta_1 = 1$ and $\beta_2 = 0.84$) identical to theirs. We explored the effect of introducing a second state variable by comparison with the single state variable case (correponding to the limit $\rho = 1$). In most of the simulations, we followed *Rice and Tse* [1986] to use $n = 10$, corresponding to cutoff of the velocity dependence at a velocity of $2.2 \times 10^4 V_*$. However, we will also show simulations in which n was varied from 0.4 to 10 to investigate the variation of dynamical behavior with the magnitude of the cutoff velocity.

Once the friction constitutive parameters have been specified, the dynamics of the spring-slider system depends only on the initial conditions and 3 parameters characterizing the spring-slider system, namely the normalized mass M, the normalized stiffness κ, and the normalized load point velocity v_o (equation (i) of Appendix). Our simulations have shown that the transient dynamical behavior is sensitively dependent on the initial conditions. However, if the load point velocity is maintained constant, then the long term dynamical behavior (involving stable sliding, self-sustained oscillations, period doubling bifurcations or chaotic oscillations) is independent of the initial conditions. It depends only on the 3 normalized parameters (M, κ, and v_o). In the phase portrait, the trajectory either spirals towards a point attraction or a limit cycle (Figure 6). This phenomenon is referred to as "Hopf bifurcation" in nonlinear dynamics [*Thompson and Stewart*, 1986], and the onset of the bifurcation corresponds to the transition from stable sliding to limit cycle oscillations.

The critical condition for the onset of Hopf bifurcation in our spring-slider system can be derived analytically from linear stability considerations. The mathematical details are provided in the Appendix. Although our general result (equation xi) would reduce to formulae previously obtained by *Ruina* [1983], *Gu et al.* [1984] and *Horowitz* [1988] for simplified cases without high-speed cutoff of velocity dependence, it is qualitatively different from the previous results in one important respect. Other than the stiffness and the friction constitutive parameters, the condition for the onset of Hopf bifurcation in our model depends also on the load point velocity because of the incorporation of high-speed cutoff. Whereas previous studies using friction relation without high-speed cutoff can come up with critical stiffness for the onset of Hopf bifurcation which depends only on the friction constitutive parameters, the critical stiffness for our system here depends not only on the friction parameters but also the load point velocity. Hence the dynamical behavior in our sytem is sensitive to variations in friction constitutive parameters (which may be induced by, for example, cumulative slip) as well as velocity perturbations. The stability boundary for the onset of Hopf bifurcation is represented by a curve in the stiffness-load point velocity space (Figure 7). Increasing load point velocity can stabilize the frictional sliding in that it may induce a transition from limit cycle oscillations to stable sliding. In this sense, the theoretical prediction from the linear stability analysis is in qualitative agreement with our experimental observations highlighted in Figures 3 and 4.

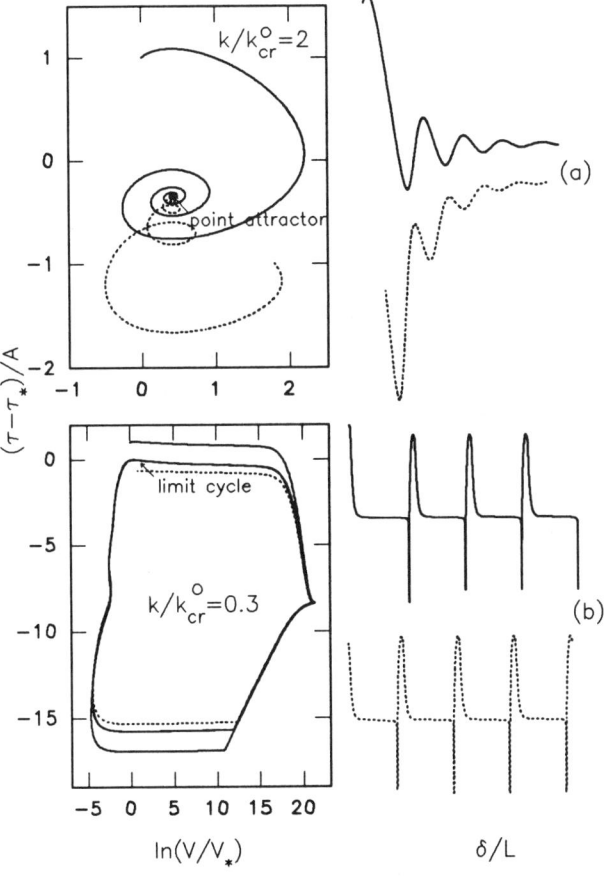

Fig. 6. Phase diagrams (left) and friction stress-slip curves (right) of a spring-slider system obeying Dieterich-Ruina rate- and state-dependent friction law with two state variables and high-speed cutoff of velocity dependence with normalized stiffness $\kappa=2\kappa_\alpha^o$ (a), and $\kappa=0.3\kappa_\alpha^o$ (b), respectively. The simulations were done with $n=10$ and normalized load point velocity $V_o/V_*=1.5$. The other friction parameters are specified in the Appendix. The normalized critical stiffness for two state variables without velocity cutoff ($n \to \infty$) is denoted by κ_α^o. The trajectory either goes to a point attractor ($\kappa=2\kappa_\alpha^o$) or a limit cycle ($\kappa=0.3\kappa_\alpha^o$) regardless of the initial conditions. The transition from a point attractor to a limit cycle corresponds to the Hopf bifurcation.

TRANSITION FROM STABLE SLIDING TO CYCLIC STICK-SLIP: NUMERICAL SIMULATIONS

Although the stability analysis detailed in the Appendix gives the critical condition for the onset of Hopf bifurcation, it does not provide any details on the nonlinear dynamical behavior of the limit cycle oscillations. The occurrence of self-sustained oscillations, period doubling bifurcations and chaotic oscillations in the transition from stable sliding to cyclic stick-slip was documented by *Gu et al.* [1984] for the two state variable case. They also concluded that period doubling and chaotic oscillations would not occur in the one state variable formulation. However, *Gu et al.* [1984] did not obtain the inertia-controlled limit cycles for cyclic stick-slip since they did not include the inertial term in their quasi-static analysis.

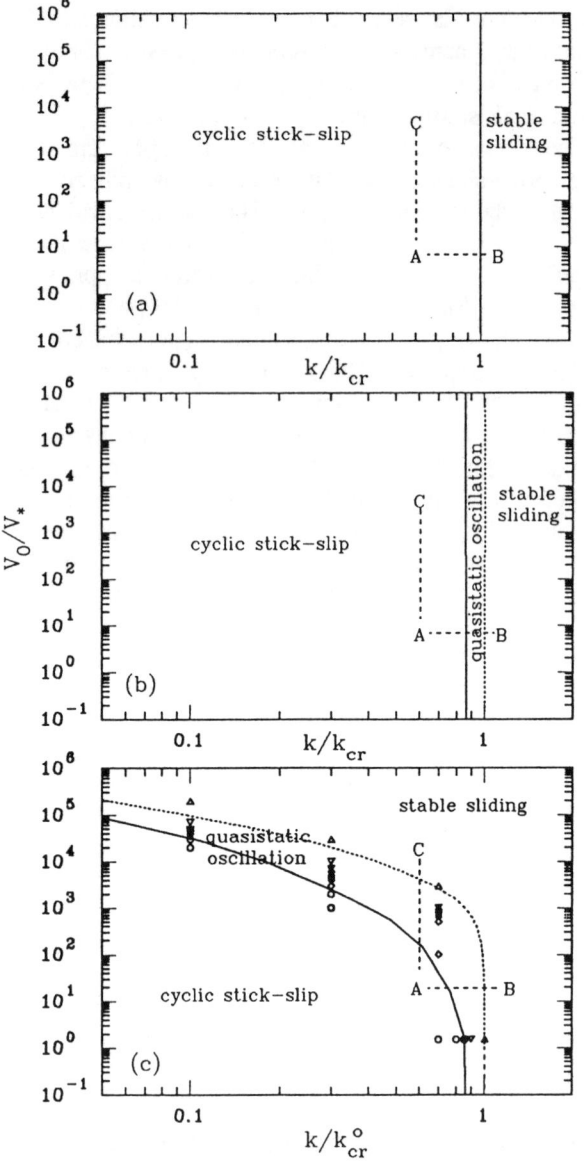

Fig. 7. Stability boundaries of a spring-slider system obeying the Dieterich-Ruina rate- and state-dependent friction law: (a) one state variable, (b) two state variables, and (c) two state variables with high-speed cutoff of velocity dependence. Horizontal axis is the stiffness k normalized with respect to either the critical stiffness k_{cr} or the critical stiffness for two state variables without velocity cutoff ($n \to \infty$) k_{cr}^o. Vertical axis is load point velocity V_o normalized with respect to a reference velocity V_*. The dotted lines mark the transition from stable sliding to quasi-static oscillation corresponding to the analytic expression for the onset of Hopf bifurcation (see Appendix), and the solid lines mark the transition from quasi-static oscillation to cyclic stick-slip determined from our numerical simulations. Here, circles represent cyclic stick-slip, diamonds period doubling, reverse triangles self-sustained periodic oscillation, squares chaotic oscillation, and triangles stable sliding.

To investigate the detailed dynamical behavior, we include inertia and used the fourth-order Runge-Kutta method to integrate the system of ordinary differential equations given by (i) in the Appendix. Details of the numerical analysis are presented in the Appendix. As we discussed before, the limit cycle oscillations depend on the 3 normalized parameters M, κ, and v_o. Following *Rice and Tse* [1986], we fixed $M = 7 \times 10^{-17}$ which according to our previous analysis [*Gu and Wong*, 1991], is a suitable value for both crustal and laboratory scale configurations.

Effect of variations in stiffness and frictional constitutive parameters

We first discuss the simulation results for two state variables without high-speed cutoff. Figure 8a illustrates the detailed dynamical behavior in the transition from stable sliding to cyclic stick-slip induced by stiffness reduction. As soon as the stiffness decreased to less than the critical value k_{cr}, self-sustained periodic oscillations were observed. As the stiffness ratio k/k_{cr} was reduced to less than 0.87, period doubling bifurcations were observed. Apparently chaotic oscillations were observed at a stiffness ratio of 0.8543, and up to this point, the oscillations were basically quasi-static. The route to chaos can be mapped out by the bifurcations in the stress drops shown in Figure 8b. As the stiffness was reduced further, the inertial term became important. For stiffness ratios between 0.85 and 0.6, several quasi-static oscillations preceded the dynamic stress drop event, and at $k/k_{cr} = 0.5$, cyclic stick-slip with a single inertia-controlled limit cycle was observed (Figure 8a). Therefore, the dominance of the inertia term caused the chaotic stress drop to remerge to a unique characteristic value at a given stiffness ratio (Figure 8c). This is analogous to the period "dedoubling" process discussed in the nonlinear dynamics literature [*Bier and Bountis*, 1984].

As far as the effect of stiffness variation is concerned, our simulation results for a model with high-speed cutoff is qualitatively similar to that without high-speed cutoff of velocity dependence. The effect of stiffness variation on the dynamical behavior for the friction law with high-speed cutoff is illustrated in Figure 9. The load point velocity was fixed (at $V_o = 1.5 V_*$), and therefore the critical stiffness (k_{cr}) for the onset of Hopf bifurcation was dependent only on the friction constitutive parameters. For stiffness greater than the critical value, the frictional sliding behavior was subcritical, ultimately leading to stable sliding. A transition from stable sliding to self-sustained oscillation was observed as soon as the stiffness k was less than k_{cr}. As the stiffness decreased further, period doubling (at $k/k_{cr} = 0.84$) and apparently chaotic

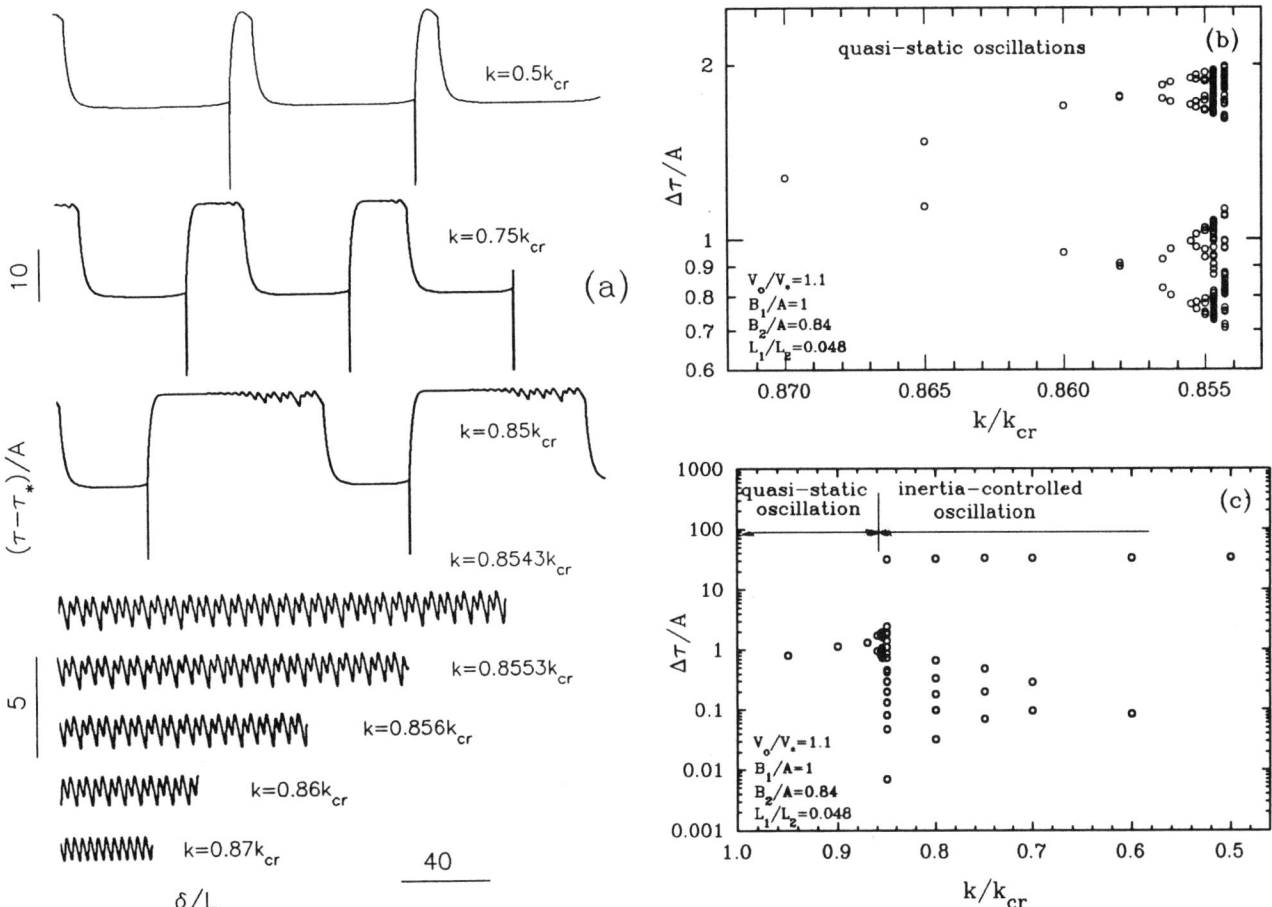

Fig. 8. (a) The effect of stiffness variation on the dynamical behavior of a spring-slider system obeying the Dieterich-Ruina rate- and state-dependent friction law with two state variables. There was no high-speed cutoff of velocity dependence. The simulations were done with normalized load point velocity $V_o/V_*=1.1$. The stiffness k is given in reference to the critical stiffness k_{cr}. Horizontal axis is normalized slip, and vertical axis is normalized frictional stress. (b) A plot of stress drop versus stiffness highlighting period doubling bifurcation in the route from stable sliding to chaos. The period doubling bifurcation pattern is clearly shown up to a period of 8 before the onset of chaos. All the oscillations were quasi-static. (c) A plot of stress drop versus stiffness from stable sliding to cyclic stick-slip. "Period dedoubling" in the inertia-controlled regime (at $k/k_{cr} < 0.85$) was evident.

oscillations (at $k/k_{cr} = 0.83$) were observed. At stiffness ratios less than 0.73, inertia-controlled limit cycle oscillations were observed.

We also investigated the effect of the cutoff velocity on the limit cycle oscillations. As the velocity cutoff exponent n was increased, the condition for the onset of Hopf bifurcation shifted to a higher load point velocity for a given stiffness (Figure 10a). Accordingly, if the load point velocity and stiffness were to be fixed while n was increased from 0.4 to 10 (and the cutoff velocity increased by 4 orders of magnitude), the sliding mode changed from self-sustained oscillation, to period doubling, and ultimately to cyclic stick-slip (Figure 10b).

For cyclic stick-slip, increasing the cutoff velocity would increase the stress drop. This is illustrated by comparing two inertia-controlled limit cycles for $n=10$ and $n=\infty$ respectively (Figure 10c). This effect is qualitatively the same as what was demonstrated by *Rice and Tse* [1986] for the one state variable case.

Effect of velocity perturbation

The major consequence of the high-speed cutoff of velocity dependence is for the frictional sliding behavior to be sensitively dependent on the load point velocity. Figure 11 highlights this dependence for load point velocity varying by 4 orders of magnitude at fixed stiffness and friction costitutive parameters. Increasing the load point velocity stabilized the frictional sliding behavior, and limit cycle oscillation mode changed from cyclic stick-slip to period doubling bifurcations to self-

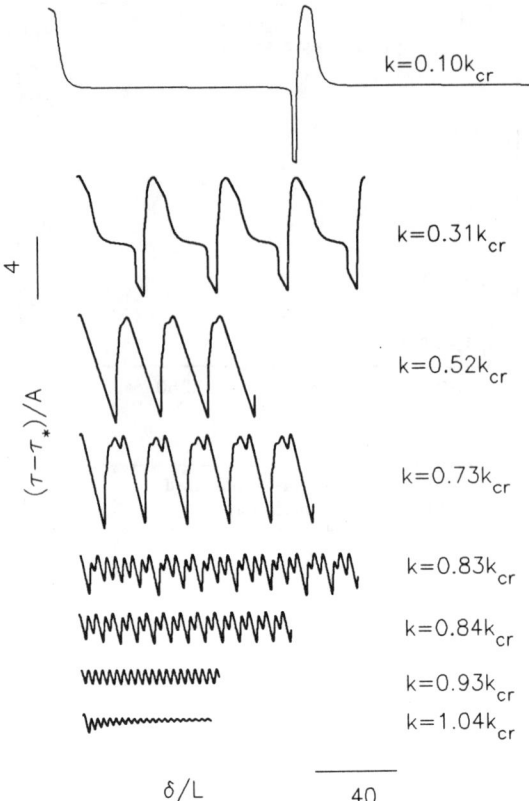

Fig. 9. The effect of stiffness variation on the dynamical behavior of a spring-slider system obeying the Dieterich-Ruina rate- and state-dependent friction law with two state variables and high-speed cutoff of velocity dependence ($n=5$). The simulations were done with normalized load point velocity $V_o/V_*=1.5$. The stiffness k is given in reference to the critical stiffness k_{cr}. Horizontal axis is normalized slip, and vertical axis is normalized frictional stress.

Fig. 10. The effect of high-speed cutoff of velocity dependence on the frictional sliding behavior of a spring-slider system. (a) Stability boundaries for different velocity cutoff $V_* e^n$. Horizontal axis is the stiffness k normalized with respect to the critical stiffness for two state variables without velocity cutoff ($n \to \infty$) k_α^o. Vertical axis is load point velocity V_o normalized with respect to the reference velocity V_*. (b) The frictional sliding behavior for different V_{cutoff} ($=V_* e^n$). The simulations were done at normalized load point velocity $V_o/V_*=1.5$, and normalized stiffness $\kappa/\kappa_\alpha^o = 0.3$ corresponding to the solid dot marked on (a). Horizontal axis is normalized slip and vertical axis is normalized friction stress. (c) Phase diagram with velocity cutoff at $n=10$ and without velocity cutoff ($n \to \infty$). The stress drop reduces significantly due to the high-speed cutoff of velocity dependence. Here, the horizontal axis is the natural logarithm of normalized sliding velocity, and the vertical axis is the normalized friction stress.

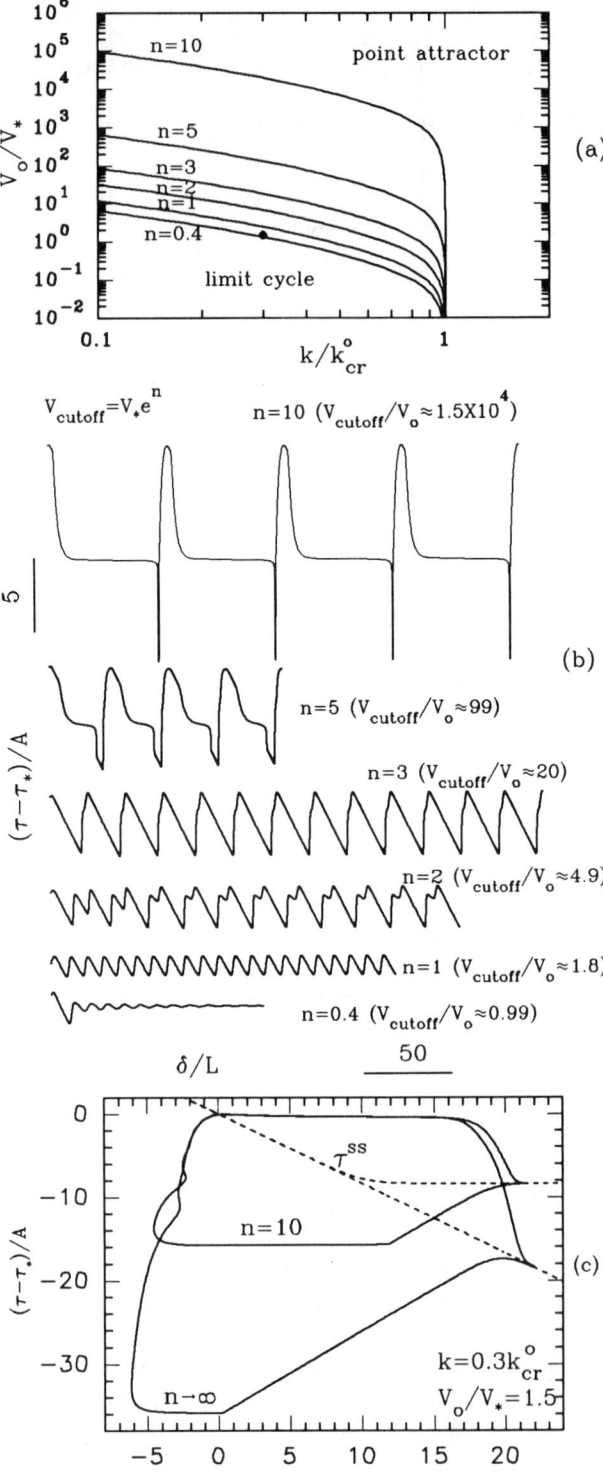

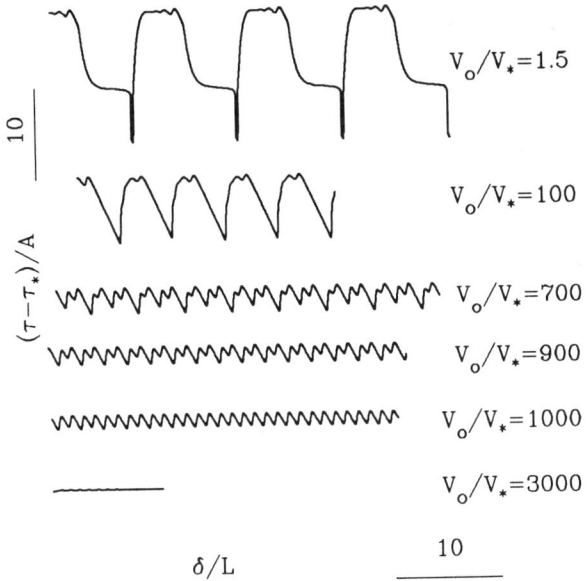

Fig. 11. The effect of load point velocity V_o on the frictional sliding behavior of a spring-slider system obeying the Dieterich-Ruina rate- and state-dependent friction law with two state variables and high-speed cutoff of velocity dependence. The simulations were done with $n=10$ and normalized stiffness $\kappa=0.7\,\kappa_\alpha^o$, where κ_α^o is the normalized critical stiffness when there is no velocity cutoff of velocity dependence ($n\to\infty$). Horizontal axis is the normalized slip and vertical axis is the normalized frictional stress.

sustained oscillation. Note that in the transitional regime, the inertia-controlled limit cycles (for $V_o/V_* = 1.5$ and 100) had several slow stress drop events preceding the major dynamic event. For $V_o/V_* = 700$, the stress drop amplitude had a period of 4 implying the occurrence of repeated bifurcation.

With additional information on the detailed nonlinear dynamical behavior from our numerical simulations, we can summarize the stability behavior by marking two transitional boundaries on the stiffness-load point velocity space (Figure 7). The onset of Hopf bifurcation according to our analytic expression (equation (xi) in the Appendix) is indicated by the dashed curve. Our numerical simulations confirm that it also marks the transition from stable sliding to self-sustained oscillations (i.e. limit cycle oscillations which are quasi-static and periodic). A second boundary marking the transition from quasi-static to inertia-controlled limit cycles is indicated by the solid curve in Figure 7. In the transitional regime embedded within the two boundaries, the limit cycle oscillations are all quasi-static and the sliding mode may be self-sustained oscillation, period doubling or chaotic.

The variation in nonlinear dynamical behavior in reponse to perturbations in stiffness ratio and load point velocity is compared for three different types of friction constitutive relation. Figure 7a (for one state variable) and Figure 7b (for two state variables) are for formulations without high-speed cutoff of the velocity dependence. Figure 7c is for two state variables with high-speed cutoff. A path from A to B corresponds to perturbation in stiffness (or friction constitutive parameters) at a fixed load point velocity. Introducing a second state variable results in an expanded transitional regime such that the transition from stable sliding (B) to cyclic stick-slip (A) goes through a path with complex limit cycle oscillations. *Wong et al.* [1992] concluded that effect of cumulative slip in destabilizing and stabilizing the sliding mode (such as the experiments shown in Figures 2 and 3) is qualitatively the same as a path between A and B in Figure 7b or 7c.

On the other hand, a path from A to C corresponds to perturbation in load point velocity keeping the stiffness and friction constitutive parameters constant. Simultaneously introducing a second state variable and high-speed cutoff of velocity dependence results in transitional boundaries dependent on both the stiffness ratio and the load point velocity, and consequently the transition from stable sliding (C) to cyclic stick-slip (A) goes through a path with complex limit cycle oscillations. The numerical simulations for a path between A and C in Figure 7c are qualitatively the same as our experimental observations shown in Figures 4 and 5.

The dynamics of the spring-slider system is such that there is a characteristic time scale associated with free oscillation of the vibration with one degree of freedom, and the Dieterich-Ruina friction law is such that there is one characteristic time (given by the ratio of the characterist slip L_i to the load point velocity) associated with each state variable [*Rice and Tse*, 1986; *Gu and Wong*, 1991]. Usually this latter set of characteristic times are significantly longer than the free oscillation time, a manifestation of the relatively long healing time of the interseismic period (of up to hundreds of years) in comparison with the dynamic rupture time (as short as several seconds). Consequently at least one state variable is necessary to model the cyclical nature of stick-slip, and incorporation of an additional state variable introduces a new characteristic time, resulting in bifurcations in the limit cycle oscillations. Furthermore, incorporation of high-speed cutoff of the velocity dependence introduces a characteristic velocity, resulting in the sliding mode being sensitive in perturbations in load point velocity.

Although we did not present the detailed results here due to space limitation, we have conducted simulations for

TABLE 1. Simulated Friction Behavior and Experimental Results

	Limit Cycle Oscillation Modes					Velocity-Dependent Stability Boundary
	Stable Sliding	Quasi-static Periodic Oscillation	Period Doubling Bifurcation	Chaotic Oscillation	Cyclic Stick-Slip	
Numerical Simulations						
1 state variable without high-speed cutoff	X	X			X	
1 state variable with high-speed cutoff	X	X			X	X
1 state variable with velocity strengthening at high speed	X	X			X	X
2 state variables without high-speed cutoff	X	X	X	X	X	
mixed 2 state variables, (1 is velocity weakening, 1 is strengthening.)	X	X	?	?	X	
2 state variables with high-speed cutoff	X	X	X	X	X	X
Experimental Observations						
Westerly granite with ultrafine quartz gouge	X	X	X	?	X	X
sawcut Tennessee sandstone	X	X	X	?	X	X
Tennessee sandstone with halite gouge	X	X	X	X	X	X

friction constitutive relations other than the specific one given by equation (2). A summary of the key observations from these numerical simulations is compiled in Table 1.

Triggering of dynamic instability by very large velocity perturbations

The above discussion has been on the long term behavior of the spring-slider system. However, it should be kept in mind that transient oscillations sensitively dependent on the initial conditions will occur before either stable sliding or limit cycle oscillations are attained. The transient behavior in response to sudden perturbations in either velocity or stress is of interest in connection with triggered seismic phenomena [*Rice and Gu*, 1983]. In the laboratory, the triggering of a dynamic instability by a large velocity perturbation of an initially stable sliding system has often been observed [e.g. *Tullis and Weeks*, 1986; *Wong and Zhao*, 1990]. Even though the stiffness may be greater than the critical value for the onset of Hopf bifurcation, a sufficient large increase in velocity or stress can induce a dynamic instability of the spring-slider system. A comprehensive quasi-static analysis of several typical initial conditions was presented by *Gu et al.* [1984] for friction laws without high-speed cutoff of velocity dependence. Here we will focus on the complication introduced by the high-speed cutoff. Specifically, the simulation results summarized in Figure 12 are for a spring-slider system initially undergoing steady state sliding (at a velocity V_o^i) and suddenly subjected to an increase of load point velocity to a new fixed value (V_o).

The response to a sudden increase in load point velocity for the Dieterich-Ruina friction law without high-speed cutoff is illustrated in Figure 12a (for the one state variable case) and Figure 12b (for the two state variable case). For a given stiffness greater than the critical value for the onset of Hopf bifurcation, there is a critical ratio of

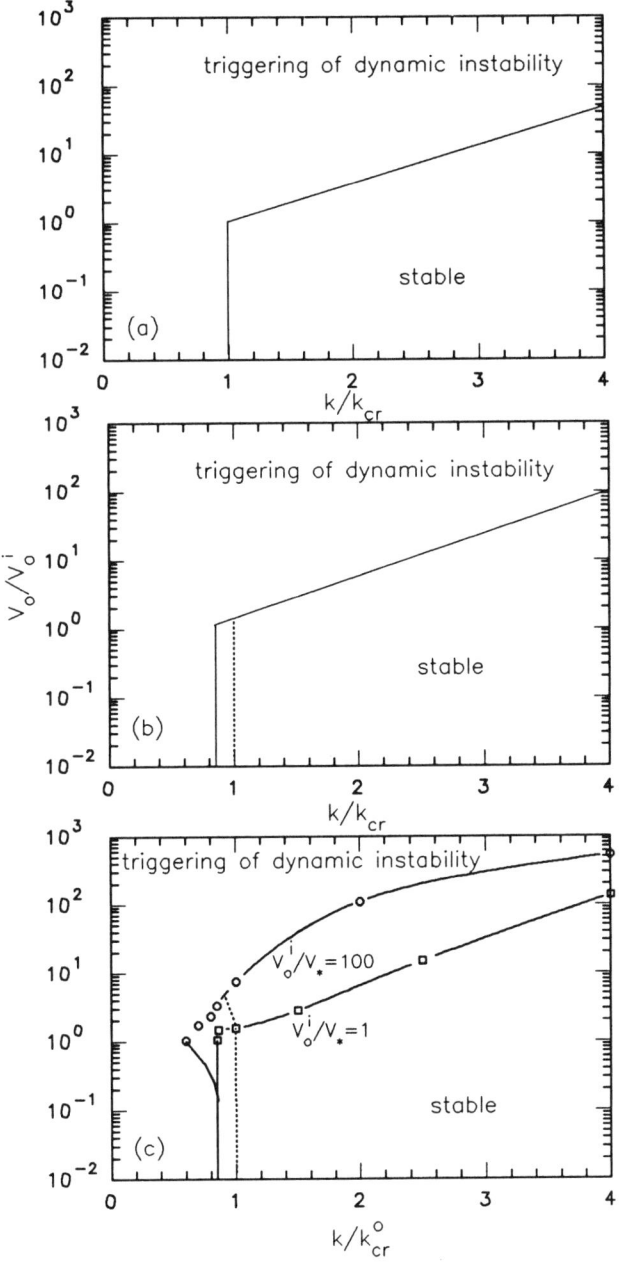

Fig. 12. Transient stability boundary of a spring-slider system obeying the Dieterich-Ruina rate- and state-dependent friction law: (a) one state variable, (b) two state variables, and (c) two state variables with high-speed cutoff of velocity dependence. For all three cases, the system is initially sliding at steady state with velocity equal to the load point velocity V_o^i, and then suddenly the load point velocity was changed to V_o. If there is no high-speed cutoff of velocity dependence, the boundary will not depend on the initial sliding velocity. However, large velocity perturbation is needed for high initial sliding velocities if there is high-speed cutoff of velocity dependence. Horizontal axis is the stiffness k normalized with respect to either the critical stiffness k_{cr} or the critical stiffness for two state variables without velocity cutoff ($n \to \infty$) k_{cr}^o. Vertical axis is load point velocity V_o normalized with respect to a reference velocity V_*.

V_o/V_o^i, above which the velocity perturbation will trigger a dynamic stress drop [Gu et al., 1984]. Subsequent to the dynamic event, the spring-slider system would undergo a number of subcritical oscillations before attaining stable sliding [Gu and Wong, 1991]. If the stiffness is such that the long term response is expected to involve quasi-static oscillations, then a sufficiently large perturbation in load point velocity can also trigger a dynamic stress drop, to be followed by quasi-static oscillations evolving ultimately to a limit cycle.

Since a characteristic velocity is intrinsic to the Dieterich-Ruina friction law with high-speed cutoff of velocity dependence, the critical condition for the triggering of a dynamic instability is dependent not only on the ratio of V_o/V_o^i, but also on the absolute values of the loading velocity. Figure 12c shows the critical condition for two loading velocities differing by 2 orders of magnitude. For a given stiffness, the velocity perturbation required to trigger a dynamic event increases with increasing initial velocity (V_o^i).

DISCUSSION

We have conducted experiments to explore the effect of cumulative slip on the frictional sliding behavior. We observed that if we maintained the normal stress and the load point velocity fixed, the accumulation of slip would first destabilize and then stabilize the sliding behavior. The slip distances over which the destabilization and stabilization processes operated seem to be significantly longer in the sawcut Tennessee sandstone than in Westerly granite samples sandwiched with ultrafine quartz gouge. This may be related to differences between the two systems with regard to the rates by which wear and Riedel shear develop in the gouge layer.

By the use of an internal load cell, we were able to systematically observe the nonlinear dynamical phenomena (including period doubling bifurcations and self-sustained oscillations) in the transition from stable sliding to cyclic stick-slip. We also investigated the variation in frictional sliding behavior in response to perturbations in the load point velocity, and we observed a general trend of stabilization by an increase of load point velocity. Apparently chaotic oscillations were also observed in Tennessee sandstone samples sandwiched with halite gouge.

Our experiments were conducted in the conventional triaxial configuration. However, similar observations have also been reported in other loading configurations. Observations of self-sustained oscillations and period doubling were respectively reported by *Scholz et al.* [1972] and *Ruina* [1980] using a direct shear apparatus. Recently, self-sustained oscillations and period doubling bifurcations were respectively observed by *Weeks and Tullis* [1985] and in Lamont [C. H. *Scholz*, personal communication, 1991] using a rotary shear apparatus. The rock types in these other studies include quartzite, granite and dolomite. We have related the complex dynamical behavior to nonlinearity intrinsic to the Dieterich-Ruina type of rate- and state-dependent friction constitutive relation. However, there is also the possibility that the complex phenomena were induced by nonlinearity of the loading system (e.g. misalignment and bending) and sample configuration (e.g. viscoelasticity of jacketing material). This type of nonlinearity can not be completely ruled out in any loading system, but one expects this type of effect to become more acute as slip is accumulated. That complex oscillations were observed at significantly different values of cumulative slip and normal stress in several rock-gouge systems suggests to us that nonlinearity of loading assembly may not be the primary cause of the nonlinear dynamical behavior in our experiments.

Key features of frictional sliding behavior observed in our numerical simulations and experiments are summarized in Table 1. In order to reproduce the full range of sliding modes observed in the laboratory and their sensitivity to velocity perturbations, we need to incorporate two state variables and high-speed cutoff of velocity dependence in the Dieterich-Ruina friction law. We arrive at this conclusion on the basis of qualitative comparison of the experimental data with numerical simulations. This conclusion can be strengthened further if we have conducted quasi-static experiments in parallel, such that we can quantitatively determine all the relevant friction constitutive parameters and input these measurements into our simulations. However, there are currently many technical problems coupled together which need to be resolved. In a system such as Westerly granite with ultrafine quartz gouge, the behavior is so unstable that it is impossible to conduct quasi-static experiments under conditions of interest. To observe the high-speed cutoff, one needs to characterize the velocity dependence over a wide range of velocities, and such a test would require slip over long distances. On the other hand, our dynamic data suggest that the friction constitutive parameters for the three rock-gouge systems vary significantly as a function of cumulative slip. It remains a formidable challenge to decouple the various factors in an experiment so as to obtain unambiguous measurements of the friction constitutive parameters.

As far as the extrapolation of laboratory derived friction law to crustal faulting is concerned, the major uncertainty remains to be the scaling of the friction parameters. The micromechanical bases of these phenomenological parameters have to be elucidated by quantitative relations between slip-induced microstructures and the friction constitutive relation. Our experiments were for three different rock-gouge systems under room temperature conditions. Recent quasi-static experiments suggest that even more complex formulations may be required to simulate behavior under hydrothermal conditions [*Blanpied et al.*, 1991; *Higgs and Chester*, 1992] relevant to the Earth's crust.

The Dieterich-Ruina friction law has begun to be used in continuum models of crustal faulting. *Tse and Rice* [1986] analyzed the earthquake cycle for a strike-slip fault with depth variation of friction parameters. An analogous analysis of dip-slip faulting was presented by *Stuart* [1988]. The spatial heterogeneity of slip and the dynamic rupture process were respectively analyzed by *Horowitz and Ruina* [1989] and *Okubo* [1989] for a homogeneous fault mode. Although their mathematical formulations differ somewhat, the friction constitutive relations used in these four studies have the common feature that they all use one state variable with high-speed cutoff of velocity dependence. If indeed two state variables with high-speed cutoff are required to model crustal faulting, then the temporal heterogeneity in a continuum model is expected to be as complex as what we have observed in the laboratory and in numerical simulation of a spring-slider system. A fundamental understanding of the dynamics of these phenomena is useful for the modeling of creep waves, slow earthquakes and periodicity of characteristic earthquakes.

Significant spatial heterogeneity may also develop in a homogeneous continuum model as was demonstrated by *Horowitz and Ruina's* [1989] study. The friction environment of a realistic seismogenic system is expected to be spatially heterogeneous due to variations in temperature, loading rate, pressure, and past tectonic activities. Many important questions related to the development of spatio-temporal complexity of slip in such a seismogenic system are still unresolved until more sophisticated continuum models are investigated [*Rice*, 1992].

To avoid the intensive computation required for a continuum model, a number of recent studies took an alternative approach using cellular automaton or spring-slider models [e.g. *Bak and Tang*, 1989; *Sornette and Sornette*, 1989; *Nakanishi*, 1990; *Carlson and Langar*, 1989; *Barriere and Turcotte*, 1991] in the spirit of *Burridge and Knopoff's* [1967] seminal work, These multiple-degree-of-freedom discrete models only involve elastic interaction among nearest neighbors. Although these models are often based on simplistic friction laws without any slip or velocity dependence, they are capable of producing complex slip patterns indicative of self-organized critical phenomena, the size distribution of which follow the Gutenberg-Richter relation. However, temporal complexity related to phenomena such as seismic quiescence and aftershocks cannot be adequately modeled unless one incorporates plasticity [*Ito and Matsuzaki*, 1990], velocity dependence [*Carlson*, 1991], memory of past sliding history [*Cao and Aki*, 1984] and dynamic overshoot [*Brown et al.*, 1991] into these discrete frictional sliding models. It seems therefore that a full consideration of the spatio-temporal complexity of the seismogenic process would still involve relatively complex interaction and friction laws, the formulation of which requires guidance from experimental rock friction.

APPENDIX

Governing Equations: Dynamic and Quasi-Static

For any system, a stable state has to be an equilibrium state. The system is stable at an equilibrium state if it comes back to the equilibrium state when a small perturbation is imposed, and it is unstable if the system moves away from the equilibrium state. We studied the stability problem by linearized analysis for a single-degree-of-freedom spring-slider system whose frictional surface obeys the Dieterich-Ruina rate- and state-dependent friction law with two state variables and high-speed cutoff of velocity dependence (equation 2). The governing equations for this system are obtained by substituting (2) in to the equation of motion (3). Using the normalization suggested by *Gu et al.* [1984], we arrived at a system of four ordinary differential equations:

$$\frac{d\phi}{dT} = e^{-\phi} \alpha / M \qquad (ia)$$

$$\frac{d\Theta_1}{dT} = -\frac{1}{2} e^{\phi} (1 + \frac{1}{\rho}) \left[\Theta_1 - (\beta_1 - \frac{1}{2}) \ln(e^{-\phi} + e^{-n}) + \frac{1}{2}\phi \right], \qquad (ib)$$

$$\frac{df}{dT} = \frac{d\phi}{dT} + \frac{d\Theta_1}{dT} - e^{\phi} \frac{1+\rho}{2} \left[f - \Theta_1 - \frac{1}{2}\phi - (\beta_1 - \frac{1}{2})\ln(e^{-\phi} + e^{-n}) \right], \qquad (ic)$$

$$\frac{d\alpha}{dT} = \kappa(v_o - e^{\phi}) - \frac{df}{dT}, \qquad (id)$$

where $\phi = \ln(\frac{V}{V_*})$, $\Theta_1 = \frac{\theta_1}{A}$, $f = \frac{(\tau - \tau_*)}{A}$, $\alpha = \frac{am}{A}$,

$T = \frac{2tV_*}{(L_1 + L_2)}$, $M = \frac{2mV_*^2}{(L_1 + L_2)A}$, $\kappa = \frac{k(L_1+L_2)}{2A}$, $v_o = \frac{V_o}{V_*}$,

$\rho = \frac{L_1}{L_2}$, $\beta_1 = \frac{B_1}{A}$, and $\beta_2 = \frac{B_2}{A}$. Using *Rice's* [1983] formulation, only one of the state variables needs to be explicitly considered in the above governing equations.

The stability analysis can be done for this system using the equations at quasi-static condition since the acceleration is very small around an equilibrium state so that the inertial term can be neglected. The normalized quasi-static equations are shown below:

$$\frac{d\phi}{dT} = \kappa v_o + e^{\phi} \left\{ -\kappa + \frac{1}{2}(\frac{1}{\rho} - \rho)\Theta_1 + \frac{1}{2}(1+\rho)f + \frac{1}{4}(\frac{1}{\rho} - \rho)\phi \right.$$
$$\left. -\frac{1}{2}\left[(1+\frac{1}{\rho})(\beta_1 - \frac{1}{2}) + (1+\rho)(\beta_2 - \frac{1}{2})\right] \ln(e^{-\phi} + e^{-n}) \right\}, \qquad (iia)$$

$$\frac{df}{dT} = \kappa(v_o - e^{\phi}), \qquad (iib)$$

$$\frac{d\Theta_1}{dT} = -\frac{1}{2}(1+\frac{1}{\rho})e^{\phi}\left[\Theta_1 + \frac{1}{2}\phi - (\beta_1 - \frac{1}{2})\ln(e^{-\phi} + e^{-n})\right]. \qquad (iic)$$

Onset of Hopf Bifurcation: Linear Stability Analysis

For such a system, there is only one equilibrium state characterized by $\frac{d\phi}{dT} = 0$, $\frac{df}{dT} = 0$, and $\frac{d\Theta_1}{dT} = 0$ at point $(\phi^o, f^o, \Theta_1^o)$, where

$$\phi^o = \ln(v_o), \qquad (iiia)$$

$$f^o = (\beta_1 + \beta_2 - 1)\ln(\frac{1}{v_o} + e^{-n}), \qquad (iiib)$$

$$\Theta_1^o = (\beta_1 - \frac{1}{2})\ln(\frac{1}{v_o} + e^{-n}) - \frac{1}{2}\ln(v_o). \qquad (iiic)$$

Physically, this state corresponds to steady state sliding

with velocity equal to the load point velocity, and friction equal to the steady state friction. Linearizing equations (ii) at the equilibrium point (iii), we get

$$\frac{d\phi}{dT} \approx \frac{1}{2}v_o\left\{\left[\frac{1}{2}(\frac{1}{\rho}-\rho)-2\kappa+\left[(1+\frac{1}{\rho})(\beta_1-\frac{1}{2})+(1+\rho)(\beta_2-\frac{1}{2})\right]\frac{1}{1+v_oe^{-n}}\right\}(\phi-\phi^o)$$
$$+\frac{1}{2}v_o(1+\rho)(f-f^o)+\frac{1}{2}v_o(\frac{1}{\rho}-\rho)(\Theta_1-\Theta_1^o),$$
(iva)

$$\frac{df}{dT} \approx -\kappa v_o(\phi-\phi^o),$$
(ivb)

$$\frac{d\Theta_1}{dT} \approx -\frac{1}{2}v_o(1+\frac{1}{\rho})\left\{\left[(\beta_1-\frac{1}{2})\frac{1}{1+v_oe^{-n}}+\frac{1}{2}\right](\phi-\phi^o)+(\Theta_1-\Theta_1^o)\right\}.$$
(ivc)

The general solutions for these linearized equations are

$$\phi = \phi^o + Ce^{rT},$$
(va)

$$f = f^o + De^{rT},$$
(vb)

$$\Theta_1 = \Theta_1^o + Ee^{rT},$$
(vc)

where C, D and E are constants. Substitute these solutions into (iv), we get

$$\left\{r-v_o\left[\frac{G-H}{2}-\kappa+G(\beta_1-\frac{1}{2})F+H(\beta_2-\frac{1}{2})F\right]\right\}C$$
$$-v_oHD-v_o(G-H)E=0,$$
(via)

$$\kappa v_o C + rD = 0,$$
(vib)

$$v_oG\left[\frac{1}{2}+(\beta_1-\frac{1}{2})F\right]C+(Gv_o+r)E=0,$$
(vic)

where $F=\frac{1}{1+v_oe^{-n}}$, $G=\frac{1}{2}(1+\frac{1}{\rho})$, and $H=\frac{1}{2}(1+\rho)$.

One trivial solution for equation (vi) is $C=D=E=0$ which corresponds to the steady state friction. If a small perturbation is superimposed on the steady state, non-zero solutions for constants C, D, and E are required. Therefore, the coefficient matrix of equation (vi) should be zero, that is

$$\begin{vmatrix} r-I & -v_oH & -v_o(G-H) \\ \kappa v_o & r & 0 \\ v_oG(\beta_1-\frac{1}{2})F+\frac{1}{2}v_oG & 0 & v_oG+r \end{vmatrix}=0,$$
(vii)

where $I=\frac{1}{2}v_o(G-H)-\kappa v_o+v_oG(\beta_1-\frac{1}{2})F+v_oH(\beta_2-\frac{1}{2})F$.

The left hand side of equation (vii) is a cubic polynomial of r,

$$A_0r^3+A_1r^2+A_2r+A_3=0,$$
(viii)

where

$$A_0 = 1,$$
(ixa)

$$A_1 = v_oG-I,$$
(ixb)

$$A_2 = -v_oGI+v_o^2H\kappa+v_o^2G(G-H)(\beta_1-\frac{1}{2})F+\frac{1}{2}v_o^2G(G-H),$$
(ixc)

$$A_3 = v_o^3HG\kappa.$$
(ixd)

From solution (v), we know that the system will come back to the steady state friction if $\text{Re}(r)<0$, and move away from the steady state if $\text{Re}(r)>0$. In other word, the steady state will be a stable state if $\text{Re}(r)<0$. The real parts of all three roots of r are smaller than zero if and only if the Routh-Hurwitz criteria are met when $A_0>0$ [Levinson and Redheffer, 1970]:

$$A_1 > 0,$$
(xa)

$$\begin{vmatrix} A_1 & A_0 \\ A_3 & A_2 \end{vmatrix} > 0,$$
(xb)

$$\begin{vmatrix} A_1 & A_0 & 0 \\ A_3 & A_2 & A_1 \\ 0 & 0 & A_3 \end{vmatrix} > 0.$$
(xc)

Solving for κ from these inequalities, we find that the conditions for stability is

$$\kappa > \frac{1}{2}\left[(1+\frac{1}{\rho})(\beta_1-\frac{1}{2})+(1+\rho)(\beta_2-\frac{1}{2})\right]\frac{1}{1+v_oe^{-n}}-\frac{(1+\rho)^2}{4\rho}=\kappa'_{cr},$$
(xia)

and

$$\kappa > \frac{1}{4\rho}\left\{\left[(\beta_1-\frac{1}{2})+\rho^2(\beta_2-\frac{1}{2})+2\rho(\beta_1+\beta_2-1)\right]\frac{1}{1+v_o e^{-n}}-\frac{(1+\rho^2)}{2}\right.$$
$$\left.+\sqrt{\left[\frac{1}{1+v_o e^{-n}}\left((\beta_1-\frac{1}{2})+\rho^2(\beta_2-\frac{1}{2})\right)-\frac{1+\rho^2}{2}\right]^2+\frac{4\rho^2(\beta_1+\beta_2-1)}{1+v_o e^{-n}}}\right\}=\kappa_{cr}^+ \quad \text{(xib)}$$

or

$$\kappa > \frac{1}{2}\left[(1+\frac{1}{\rho})(\beta_1-\frac{1}{2})+(1+\rho)(\beta_2-\frac{1}{2})\right]\frac{1}{1+v_o e^{-n}}-\frac{(1+\rho)^2}{4\rho}=\kappa_{cr}', \quad \text{(xia)}$$

and

$$\kappa < \frac{1}{4\rho}\left\{\left[(\beta_1-\frac{1}{2})+\rho^2(\beta_2-\frac{1}{2})+2\rho(\beta_1+\beta_2-1)\right]\frac{1}{1+v_o e^{-n}}-\frac{(1+\rho^2)}{2}\right.$$
$$\left.-\sqrt{\left[\frac{1}{1+v_o e^{-n}}\left((\beta_1-\frac{1}{2})+\rho^2(\beta_2-\frac{1}{2})\right)-\frac{1+\rho^2}{2}\right]^2+\frac{4\rho^2(\beta_1+\beta_2-1)}{1+v_o e^{-n}}}\right\}=\kappa_{cr}^- \quad \text{(xic)}$$

Since $\kappa_{cr}^+ \geq \kappa_{cr}^-$, the normalized critical stiffness is

$$\kappa_{cr} = \frac{k_{cr}(L_1+L_2)}{2A} = \max(\kappa_{cr}', \kappa_{cr}^+). \quad \text{(xii)}$$

In general, κ_{cr} will increase with the increasing n and decreasing v_o. Either k_{cr}' or k_{cr}^+ could be the desired critical value depending on the relative value of ρ, β_1, β_2, n and v_o. However, with one state variable ($\beta=\beta_1+\beta_2$, $L_1=L_2$), $\kappa_{cr}^+ > \kappa_{cr}'$, and therefore, $\kappa_{cr} = \kappa_{cr}^+ = (\beta-1)/(1+v_o e^{-n})$, or explicitly, $k_{cr} = (B-A)/\{L[1+(V_o/V_*)e^{-n}]\}$.

If there is no high-speed cutoff of velocity dependence ($n \to \infty$), the critical stiffness will be independent of load point velocity V_o. With one state variable, the stiffness reduces to $k_{cr} = (B-A)/L$, which was derived by *Ruina* [1983]. With two state variables, equation (xi) reduces to *Horowitz*'s [1988] result if $n \to \infty$, and we will denote $\kappa_{cr}^+(n \to \infty)$ by $\kappa_{cr}^o(\beta_1, \beta_2, \rho)$:

$$\kappa_{cr}^o = \left\{\beta_1-1+\rho^2(\beta_2-1)+2\rho(\beta_1+\beta_2-1)\right.$$
$$\left.+\sqrt{[\beta_1-1+\rho^2(\beta_2-1)]^2+4\rho^2(\beta_1+\beta_2-1)}\right\}/4\rho \quad \text{(xiii)}$$

For most of our simulations shown in this paper, we chose $\rho=0.048$, $\beta_1=1$, and $\beta_2=0.84$ [*Gu et al.*, 1984], and $n=10$ [*Rice and Tse*, 1986]. For this choice of parameters, $k_{rc} = k_{cr}^+$, and it can be shown that the angular frequency of the self-sustained oscillation should be given by

$$\omega = \frac{V_o(1+\rho)}{(L_1+L_2)\sqrt{\rho}}\sqrt{2\kappa_{cr}-(\beta_1+\beta_2-1)\frac{1}{1+v_o e^{-n}}}, \quad \text{(xiv)}$$

which reduces to the result of *Gu et al.* [1984] if $n \to \infty$.

Different from the analysis done before by *Rice and Ruina* [1983], *Gu et al.* [1984] and *Horowitz* [1988], we linearized the governing equation at an equilibrium state directly without using the Laplace transform. In a linear stability analysis, we can only analyze the properties near a steady sliding state. That $\kappa = \kappa_{cr}$ corresponds to the onset of Hopf bifurcation where the trajectory of the system changes from a point tractor to a limit cycle was confirmed by our numerical simulations.

Numerical Simulations

The fourth-order Runge-Kutta method was used to integrate equations (i) or (ii). The criterion for us to change from equations (ii) to (i) is that α became bigger than 1% of f, and to change from equations (i) to (ii) was that $d\alpha/dT$ became smaller than at least 1% of df/dT. The time-step was allowed to evolve with time. We did not allow $d\alpha/dT$, df/dT, $d\phi/dT$, and $d\Theta_1/dT$ to vary by more than 10% for each time step except when one of the four derivatives was close to zero. To ensure the effectiveness of the calculation, we also kept the change of these four derivatives bigger than 0.1%. We have also checked several cases with finer time steps, for example, keeping the change $d\alpha/dT$, df/dT, $d\phi/dT$, and $d\Theta_1/dT$ less than 1% and bigger than 0.01% in each time step. There was no observable difference between them except for the chaotic oscillations. In such a case, the system would still oscillate chaotically, but the details could be very different.

In our simulations, we chose initial conditions as $\phi=0$, $f=0$, $\Theta_1=0$, and $\alpha=0$. However, we did check other arbitrarily chosen initial conditions. Except in the chaotic regime, the initial conditions did not affect the long-term behavior at all, but the transient behaviors could be totally different. In the chaotic regime, different initial conditions would still keep the system in the chaotic regime, but the pattern could not be repeated.

Acknowlegements. The Tennessee sandstone was kindly furnished by Gene Scott. This research was supported by the U. S. Geological Survey through grant 14-08-0001-G1807.

REFERENCES

Bak, P. and C. Tang, Earthquakes as a self-organized critical phenomenon, *J. Geophys. Res.* **94**, 15635-15637, 1989.

Barriere, B., and D. L. Turcotte, A scale-invariant cellular-automata model for distributed seismicity, *Geophys. Res. Lett.*, **18**, 2011-2014, 1991.

Biegel, R. L., C. G. Sammis, and J. H. Dieterich, The frictional properties of a simulated gouge with fractal particle distribution, *J. Struct. Geol.*, **11**, 827-846, 1989.

Bier, M. and T. C. Bountis, Remerging Feigenbaum trees in dynamical systems, *Physics Letters A*, **104**, 239-244, 1984.

Blanpied, M. L., Friction Constitutive Behavior and Textural Evolution of Experimental Faults in Granite, Ph. D. thesis, Brown University, 152 pp., 1989.

Blanpied, M. L., T. E. Tullis, and J. D. Weeks, Frictional behavior of granite at low and high sliding velocities, *Geophys. Res. Lett.*, **14**, 554-557, 1987.

Blanpied, M. L., D. A. Lockner, J. D. and Byerlee, Fault stability inferred from granite sliding experiments at hydrothermal conditions, *Geophys. Res. Let.*, **18**, 609-612, 1991.

Brace, W. F., and J. D. Byerlee, Stick-slip as a mechanism for earthquakes, *Science*, **153**, 990-992, 1966.

Brown, C. H. Scholz, and J. B. Rundle, A simplified spring-block model of earthquakes, *Geophys. Res. Lett.*, **18**, 215-218, 1991.

Burridge, R., and L. Knopoff, Model and theoretical seismicity, *Bull. Seismol. Soc. Am.*, **57**, 341-371, 1967.

Byerlee, J. D., Frictional characteristics of granite under high confining pressure, *J Geophys. Res.*, **72**, 3639-3648, 1967.

Byerlee, J. D., A review of rock mechanics studies in the United States pertinent to earthquake prediction, *Pure Appl. Geophys.*. **116**, 586-602, 1978.

Byerlee, J. D. & R. Summers, Stable sliding preceding stick-slip on fault surfaces in granite at high pressure, *Pure Appl. Geophys.*, **113**, 63-68, 1975.

Cao, T., and K. Aki, Seismicity simulation with a rate and state dependent friction law, *Pure Appl. Geophys.*, **124**, 487-513, 1986.

Carlson, L. M., Time intervals between characteristic earthquakes and correlations with smaller events: an analysis based on a mechanical model of a fault, *J. Geophys. Res.*, **96**, 4255-4267, 1991.

Carlson, J. M., and J. S. Langar, Properties of earthquakes generated by fault dynamics, *Phys. Rev. Let.*, **62**, 2632-2635, 1989.

Chester, F. M., and N. G. Higgs, Multi-mechanism friction constitutive model for ultra-fine quartz gouge at hypocentral conditions, *J. Geophys. Res.*, in press, 1992.

Chester, F. M. & J. M. Logan, Frictional faulting in polycrystalline halite: correlation of microstructure, mechanisms of slip, and constitutive behavior. in *The Brittle-Ductile Transition in Rocks*, Edited by A. G. Duba, W. B. Durham, J. W. Handin and H. F. Wang, AGU, Washington, D. C., Geophysical Monograph **56**, 49-66, 1990.

Cox, S. J. D., Velocity-dependent friction in a large direct shear experiment on gabbro, in *Deformation Mechanisms, Rheology and Tectonics*, edited by R. J. Knipe and E. H. Rutter, Geological Society Special Publication No. 54, 63-70, 1990.

Dieterich, J. H., Time-dependent friction in rock, *J. Geophys. Res.*, **77**, 3690-3697, 1972.

Dieterich, J. H., Time dependent friction and the mechanics of stick-slip, *Pure Appl. Geophys.*, **116**, 790-806, 1978.

Dieterich, J. H., Modeling of rock friction 1: Experimental results and constitutive equation, *J. Geophys. Res.*, **84**, 2161-2168, 1979.

Dieterich, J. H., Constitutive properties of faults with simulated gouge, in *Mechanical Behavior of Crustal Rocks*, Edited by N. L. Carter, M. Friedman, J. M. Logan, and D. W. Stearns, 102-120, AGU monograph **24**, 1981.

Dieterich, J. H., A model for the nucleation of earthquake slip, in *Earthquake Source Mechanics*, edited by Das et al., Am. Geophys. Union Geophys. Monogr., **37**, 37-47, 1986.

Engelder, J. T., J. M. Logan, and J. Handin, The sliding characteristics of sandstone on quartz fault-gouge, *Pure Appl. Geophys.*, **113**, 68-86, 1975.

Gu, J-c., J. R. Rice, A. L. Ruina, and S. T. Tse, Slip motion and stability of a single degree of freedom elastic system with rate and state dependent friction, *J. Mech. Phys. Solids*, **32**, 167-196, 1984.

Gu, Y. & T.-f. Wong, Effects of loading velocity, stiffness, and inertia on the dynamics of a single degree of freedom spring-slider system, *J. Geophys. Res.*, **96**, 21,677-21,691, 1991.

Gu, Y., and T.-f. Wong, The transition from stable sliding to cyclic stick-slip: effect of cumulative slip and load point velocity on the nonlinear dynamical behavior in three rock-gouge systems, in *Rock Mechanics Proceedings of the 33rd U. S. Symposium*, edited by J. R. Tillerson and W. R. Wawersik, 151-158, 1992.

Horowitz, F. G., Mixed state variable friction laws: Some implications for experiments and a stability analysis, *Geophys. Res. Lett.*, **15**, 1243-1246, 1988.

Horowitz, F. G. and A. Ruina, Slip patterns in a spatially homogeneous fault model, *J. Geophys. Res.*, **94**, 10279-10298, 1989.

Ito, K. and M. Matsuzaki, Earthquakes as self-organized critical phenomena, *J. Geophys. Res.*, **95**, 6853-6860, 1990.

Jaeger, J. C., and N. G. W. Cook, Friction in granular materials, in *Proc. of the Civil Engineering Materials Conference Southampton, Part I*, edited by A. M. Te'eni, 257-66, 1971.

Jones, D. E., F. E. Kennedy, and E. M. Schulson, The kinetic friction of saline ice against itself at low sliding velocities, *Annals of Glaciology*, **15**, 242-246, 1991.

Kanamori, H., and C. R. Allen, Earthquake repeat time and average stress drop, in *Earthquake Source Mechanics*, edited by Das et al., Am. Geophys. Union Geophys. Monogr., 37, 227-235, 1986.

Levinson, N., and R. M. Redheffer, *Complex Variables*, Holden-Day, San Francisco, 1970.

Linker, M. F., and J. H. Dieterich, Effects of variable normal stress on rock friction: observations and constitutive equations, *J. Geophys. Res.*, 4923-4940, 1992.

Madariaga, R., A. Cochard, M. Koller, and M. Bonnet, Fracture mechanical model of a heterogeneous earthquake fault, AGU abstracts, EOS, *Trans. Am. Geophys. Union*, 72, 325, 1991.

Marone, C., C. H. Scholz, and R. Bilham, On the mechanics of earthquake afterslip, *J. Geophys. Res.*, 96, 8441-8452, 1991.

Nakanishi, H., Cellular-automaton model of earthquakes with deterministic dynamics, *Phys. Rev. A*, 41, 7086-7089, 1990.

Ohnaka, M., Experimental studies of stick-slip and their application to the earthquake source mechanism, *J. Phys. Earth*, 21, 285-303, 1973a.

Okubo, P., Dynamic rupture modeling with laboratory-based constitutive relations, *J. Geophys. Res.*, 94, 12321-12335, 1989.

Okubo, P. G., and J. H. Dieterich, State variable fault constitutive relations for dynamic slip, in *Earthquake Source Mechanics*, Am. Geophys., edited by S. Das, J. Boatwright, and C. H. Scholz, Union monograph 37, 25-35, 1986.

Paterson, M. S., *Experimental Rock Deformation—the Brittle Field*, Speringer-Verlag, New York, 1978.

Rabinowicz, E., The intrinsic variables affecting the stick-slip process, *Proc. Phys. Soc.*, 71, 668-675, 1958.

Rayleigh, J. W. S., *The Theory of Sound*, Reprinted by Dover, New York, 1986.

Reinen, L. A., J. D. Weeks, and T. E. Tullis, The frictional behavior of serpentine: implication for aseismic creep on shallow crustal faults, *Geophys. Res. Lett.*, 18, 1921-1924, 1991.

Rice, J. R., Constitutive relations for fault slip and earthquake instabilities, *Pure Appl. Geophys.*, 121, 443-475, 1983.

Rice, J. R., Spatio-temporal complexity of slip on a fault, Harvard University, Division of Applied Sciences, Report MECH-195, pp 37, 1992.

Rice, J. R. and J.-c. Gu, Earthquake aftereffects and triggered seismic phenomena, *Pure Appl. Geophys.*, 121, 187-219, 1983.

Rice, J. R., and A. L. Ruina, Stability of steady frictional slipping, *J. App. Mech.*, 50, 343-349, 1983.

Rice, J. R., and S. T. Tse, Dynamic motion of a single degree of freedom system following a rate and state dependent friction law, *J. Geophys. Res.*, 91, 521-530, 1986.

Richardson, R. S. H. and H. Nolle, Surface friction under time-dependent loads, *Wear*, 37, 87-101, 1976.

Ruina, A. L., *Friction Laws and Instabilities: a Quasi-static Analysis of some Dry Frictional Behavior*, PhD dissertation, Division of Engineering, Brown University, 1980.

Ruina, A. L., Slip instabilities and state variable friction laws, *J. Geophys. Res.*, 88, 10359-10370, 1983.

Scholz, C. H., *The Mechanics of Earthquakes and Faulting*, Cambridge University Press, Cambridge, 439 pp., 1990.

Scholz, C. H., P. Molnar, and T. Johnson, Detailed study of frictional sliding of granite and implications for the earthquake mechanism, *J. Geophys. Res.*, 77, 6392-6406, 1972.

Scott, T. E. & K. C. Nielsen, The effect of porosity on the brittle-ductile transition in sandstones, *J. Geophys. Res.*, 96, 405-414, 1991.

Shimamoto, T. and J. M. Logan, Velocity-dependent behaviors of simulated halite shear zones: An argument for silicates, in *Earthquake Source Mechanics*, edited by S. Das et al., Am. Geophys. Union Monogr., 37, 49-63, 1986.

Sornette, A., and D. Sornette, Self-organized criticality and earthquakes, *Europhys. Lett.*, 9, 197-202, 1989.

Stesky, R. M., Rock friction - effect of confining pressure, temperature, and pore pressure, *Pure Appl. Geophys.*, 116, 690-704, 1978.

Stoker, J. J., *Nonlinear Vibrations*, Wiley New York, 1950.

Stuart, W. D., Forecast model for great earthquakes at the Nankai Trough subduction zone, *Pure Appl. Geophys.*, 126, 619-641, 1988.

Teufel, L. W., and J. M. Logan, Effect of displacement rate on the real area of contact and temperatures generated during frictional sliding of Tennessee sandstone, *Pure Appl. Geophys.*, 116, 840-865, 1978.

Thompson, J. M. T. and H. B. Stewart, *Nonlinear Dynamics and Chaos*, John Wiley and Sons, New York, 376 pp. 1986.

Tse, S. T., and J. R. Rice, Crustal earthquake instability in relation to the depth variation of frictional slip properties, *J. Geophys. Res.*, 91, 9452-9472, 1986.

Tullis, T. E., Friction and faulting, editor's note, *Pure Appl. Geophys.*, 124, 375-382, 1986.

Tullis, T. E., and J. D. Weeks, Constitutive behavior and stability of frictional sliding of granite, *Pure Appl. Geophys.*, 124, 383-414, 1986.

Weeks, J. D. and Tullis, T. E., Frictional sliding of dolomite: a variation in constitutive behavior, *J. Geophys. Res.*, 90, 7821-7826, 1985.

Wesnousky, S. G., Seismological and structural evolution of strike-slip faults, *Nature* 335, 340-342., 1988.

Wesnousky, S. G., Seismicity as a function of cumulative geologic offset: some observations from southern California, *Bull. Seismo. Soc. Am.*, 80, 1374-1381, 1990.

Wong, T.-f., and Y. Zhao, Effects of load point velocity on frictional instability behavior, *Tectonophysics*, 175, 177-195, 1990.

Wong, T.-f., Y. Gu, T. Yanagidani and Y. Zhao, Stabilization of faulting by cumulative slip, in *Fault Mechanics and Transport Properties of Rock*, Edited by B. Evans and T.-f. Wong, Academic Press, 109-133, 1992.

Y. Gu and T.-f. Wong, Department of Earth and Space Sciences, State University of New York at Stony Brook, Stony Brook, NY 11794-2100.

Is the Dynamics of the Lithosphere Chaotic?

Q. Li and E. Nyland

Department of Physics, University of Alberta, Edmonton, Canada

A region of the lithosphere might be considered a non-linear deterministic system described by as yet unknown differential equations. Even though the governing equations are unknown, a lithospheric phase portrait for such a region can be reconstructed from a single observable such as an earthquake catalog. Using this point of view the system can be analyzed by methods which have developed from the study of non-linear dynamic systems. The low dimensionality of an attractor, between 3 and 4, obtained in the analysis of seismicity data for the west coast of Canada supports the argument that the apparent disorder in the dynamics of the lithosphere, in this area at least is chaotic rather than stochastic.

INTRODUCTION

Since earthquake related stress and strength fields of the lithosphere are usually inaccessible to direct measurement, our understanding of the mechanism of earthquake occurrence is primarily based on the results of non-linear models of the failure process in rock. One such model (Burridge and Knopoff 1967) was modified to include more complicated friction laws (Dieterich, 1978, 1979; Ruina, 1983; Rudnicki, 1988). Another considers the instability that might result from an explosive increase of temperature in faults (Griggs and Baker, 1969; Ogawa, 1987). In these models a single fault is generally considered in isolation from the others for mathematical simplicity. The onset of instability is interpreted as an earthquake.

Seismicity in a more general sense is probably evidence of a variety of complex causes. Long term behavior of seismicity appears random but it reflects integral characteristics of the lithosphere dynamics and encompasses all the complexities of faulting. The physics of failures whose characteristic size is much smaller than the characteristic dimensions of geologic structure must be quite different from that of large events with characteristic sizes similar to that of the local geologic structure. This suggests that the physics of small earthquakes may be very different from that of large ones, but it does not suggest that the processes that trigger small earthquakes are decoupled from those that trigger large ones. This argument is a justification for seeking phenomenological evidence of underlying simplicity in what appears at first sight to be a chaotic process. Statistical analysis of the nature of earthquake occurrence shows that the dependence between events is very weak (Keilis-Borok, et al., 1971) and that the mainshocks are mainly Poissonian distributed (Gardner and Knopoff, 1974). Earthquake fault zones have a three dimensional structure that has a fractal distribution of size (Kagan and Knopoff, 1980; Kagan, 1981a,b).

Even though the symptoms of strong earthquake occurrence differ from case to case, general premonitory seismic patterns prior to strong earthquakes have been observed (Keilis-Borok et al., 1988; Keilis-Borok, 1990). This indicates that gross integrated characteristics of lithosphere dynamics exist. They are similar in different regions, independent of the tectonic environment and the level of seismicity and could be associated with parameters of non-linear processes. Several studies into the possibility of chaotic earthquake occurrence (Huang and Turcotte, 1990; Carlson and Langer, 1989; Li, 1991) showed that a complicated earthquake sequence could result from the interactions between different fault segments. Such a view of fault models suggests we should treat the lithosphere as a non-linear dynamic system that receives energy in a relatively smooth, broadly spread way. Such energy is acquired by widely separated regions in a time correlated way and as a result phenomena in widely separated regions can show coupling, not because of interaction energies or self organized criticality but rather because they are driven by a common forcing function.

We ask whether evidence of such a phenomenon exists in the earthquake catalog for a portion of Western Canada. We take the view that the methods of non-linear science should be useful in this area but that the applicability of specific tools should be driven by observational knowledge. In particular we attempt to avoid, as much as possible, the use of overly detailed models of the physics behind earthquake occurrence. Our only significant hypothesis is the notion that the seismicity over one fairly large area can be described as the result of a single non-linear dynamic system.

RECONSTRUCTION OF LITHOSPHERE DYNAMICS FROM THE EARTHQUAKE CATALOG

The dynamics of the lithosphere is described by a complicated system of partial differential equations. A practical way to reduce such partial differential equations to a group of ordinary differential equations is to truncate their series expansion in the way Lorenz truncated the non-linear Navier-Stokes equations for the atmosphere (Lorenz, 1963) but unfortunately the correct form of the equations that describe lithosphere dynamics is not as well understood as the form of the equations that describe atmospheric dynamics. An alternative way to discretize the dynamics of the lithosphere is to divide a region of lithosphere into N small domains. If each patch has k state variables, the system is described by kN coupled ordinary differential equations, the state of the system at a given time will correspond to a point in a kN dimensional phase space, and the evolution of the system is described by a trajectory in this space. If there exists a low-dimensional attractor the dynamics of the lithosphere should be treated as that of a low dimensional deterministic system.

Of course we cannot expect to construct the phase space of the entire dynamics of the lithosphere. Suppose however that we know a set of points in the phase space and we choose these points to be those at which earthquakes occur. On physical grounds these failure regions must be localized and continuous; the well known example in a two-dimensional phase space with axes determined by shear and compressional stress is the Mohr Coulomb failure curve. A similar hypersurface must exist for systems with more degrees of freedom and the points where a system phase curve crosses this hypersurface are points where earthquakes occur. We focus here on the nature of this more general sub-space which we interpret as the part of the phase space in which earthquakes occur. It must have fewer dimensions than the lithospheric phase space for the criterion for failure of the quasi-static approximations has to be a constraint among the physical parameters and the sub-space should be connected on the grounds that in at least some direction an increase in one of the parameters that defines failure should always lead to failure.

For the study of lithosphere dynamics, we can imagine selecting this sub-space as a special surface of section in the high-dimensional phase space. The lithosphere states on the surface correspond to the states when an earthquake will occur and if we have complete knowledge of the lithosphere state at one of them the others are in principle determinable. In practice the observational errors limit this predictability. The time of intercept of the trajectory by the surface corresponds to the time of on-set of an earthquake. The time between the instant the trajectory leaves the surface and its next return corresponds to the interval between successive events. A collection of the points that appear on the surface correspond to an earthquake catalog. We can, therefore, consider the earthquake catalog as an experimental measure of the map of the lithospheric phase space trajectory. If we study only the time interval between events we are projecting the higher dimensional dynamics on a single variable. The degrees of freedom left out can be determined by considering a longer history of measurements. The earthquake catalog is now viewed as an output time series of a deterministic system, and the data analysis techniques developed recently for studying such systems are applicable.

If we restrict our attention to the dynamics of a finite dimensional attractor, information can be retrieved from such history data. Packard, et al. (1980) and Takens (1981) have proved the existence of an embedding from an m-dimensional manifold to a n-dimensional Euclidean space R_n defined by $\Phi(x)=(\varpi(x), \varpi(\phi_1(x)),..., \varpi(\phi_n(x)))$ when n>=2m+1, where ϕ_t is the flow of the system, and ϖ a smooth function on the manifold. This forms the basis of reconstruction techniques for phase portraits from time series measurements in experimental domains.

In practice it is necessary to relate this embedding theorem to a time series of measurements made on the system. One method of reconstructing a phase portrait from a time series is to create more signals from a single one by using time delay (Grassberger and Procaccia 1983, Roux et al. 1983). Consider a time series {B(t)} and time delay T. Suppose we create a state vector x(t) by assigning coordinates $\{x_1, x_2, ...x_n\}$ by {B(t), B(t+T), ..., B(t+(n-1)T)}. If the dynamics takes place on an attractor of dimension D, then a necessary condition for determinism is n>= D. An n-dimensional phase portrait constructed from {B(t), B(t+T), ..., B(t+(n-1)T)} will have the same properties (topologically) as one constructed from measurements of m independent variables, if n>=2m+1 where m is the dimension of the attractor. We suggest that in the case of the dynamics of the lithosphere, aspects of the portrait of the phase space trajectory can be reconstructed from the earthquake catalogs.

If the phase space is partitioned into cells with diameter ε the probability that the phase space trajectory falls into the i-th cell is P_i.

$$\text{Let } D_f = \lim_{\varepsilon \to 0} \frac{1}{f-1} \frac{\log \sum_{i=1}^{N(\varepsilon)} P_i^f}{\log \varepsilon} \qquad f = 0, 1, 2, ...$$

When f = 0 and 1, we obtain the Hausdorff dimension and information dimension respectively. When f = 2, $\log \sum_{i=1}^{N(\varepsilon)} P_i^2$ is the probability that two points of the attractor lie within a cell. This is also the probability that two points of the attractor are separated by a distance smaller than ε and becomes

$$\lim_{N \to \infty} \frac{1}{N^2} \{\text{number of pairs (i,j) whose distance } |s_i-s_j| \text{ is less than } \varepsilon\}.$$

This correlation integral C(ε) measures the spatial correlation of the points that lie on an attractor. $D_2 = \lim_{\varepsilon \to 0} \frac{\log C(\varepsilon)}{\log \varepsilon}$ is the correlation dimension whose numerical calculation is relatively easy and straight forward. The correlation integral C(ε) can be calculated from a single experimental measurement by applying the method of delays to the time series and indicates the degree of randomness of a physical system.

EARTHQUAKE ANALYSIS

We calculated D_2 for the west coast of Canada. In the area we studied, from 118°W to 130°W and from 45°N to 51°N, we merged the seismicity catalog of the Geological Survey of Canada with the catalog from the National Earthquake Information Service of the U.S.A. Particularly for the earlier years, the Canadian catalog contained many significant earthquakes as far south as 45°N. The duplicate events were removed manually. Usually NEIS data were

TABLE 1. AfterShock Windows

M	R(km)	T(days)
2.5	29	6
3.0	33	11.5
3.5	39	22
4.0	45	42
4.5	52	83
5.0	60	150
5.5	70	290
6.0	81	510
6.5	91	790
7.0	105	915
7.5	121	960
8.0	141	985

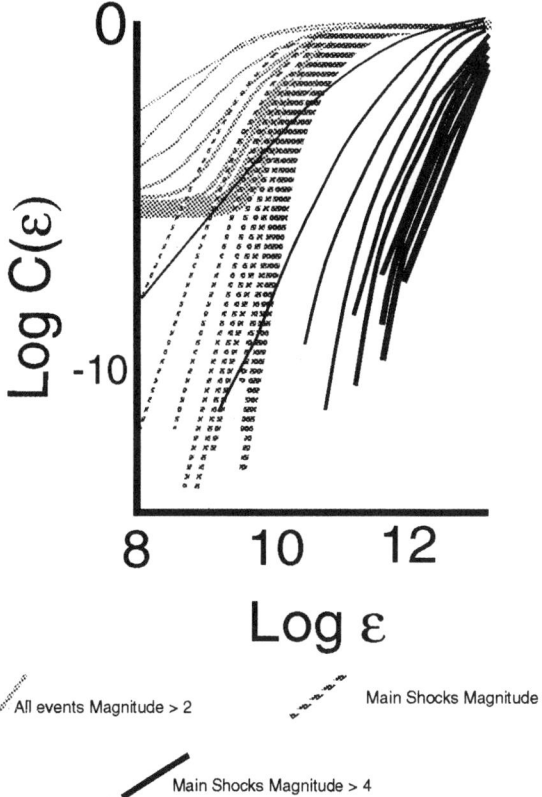

Fig.1. Log C(e) versus log e calculated for various subsets of the earthquake catalog. The plots are derived from material in Li (1991) and are for data taken in groups of 3 (the lightest line), 6, 9,... (lines of increasing width). The different line textures indicate analysis of different data sets. Each line is a spline with 5 nodes visually fit to data calculated as explained in the text

accepted in the United States and GSC data were accepted in Canada. If only one agency reported a magnitude the report of that agency was accepted. Events for which no magnitude was reported were assigned zero magnitude, and if a preferred magnitude was given it was used. For these data that is usually M_L. We also removed those events since May 1980 that were clearly related to the Mt. St. Helen's Volcanic Eruption. This catalog has been used successfully in the analysis of premonitory seismicity patterns of the area (Brown, et al., 1989), and we believe that it reflects the dynamics of the region. Analysis of earthquake catalogs may require a separation between mainshocks and aftershocks. The windows used in our analysis are listed in table 1 (Brown, et al., 1989).

The earthquake catalog has a sufficient number of small earthquakes since the early 1950's. In the period 1950 to 1985, the catalog contains 5200 seismic events, among them there are 3160 events with magnitude above 2; 360 mainshocks with magnitude above 4 and 2000 mainshocks with magnitude above 2. The mainshock catalogs remain after the aftershocks are deliberately removed from the catalog. The calculations were carried out for both mainshock catalogs and the catalog with all events in order to compare the results respectively, and see how aftershocks affect the integral dynamic evolution of the lithosphere.

We used the set of time intervals between successive events as the data series for the calculation. This choice implies a major assumption for it assumes that the dynamics of the lithosphere is consistent over the area covered by the catalogue. We believe this assumption is reasonable because we expect the area to be dominated by the process of subduction of the Juan da Fuca plate. If different mechanisms cause earthquakes in different parts of the region and if these mechanisms are not spatially coupled it will be difficult to interpret the results. An alternative data series to use in the analysis would have applied some kind of weighting to put less emphasis on earthquakes that were widely separated in space than on those that are close. We could find no rational physical basis for a choice of such a metric and in any case comment that this process stamps a much more specific model assumption on the calculation than does our process of benign neglect. We would expect high dimensionality if several unrelated mechanisms were active in this case. Low dimensionality does not prove spatial correlation, it simply says that the underlying mechanism for earthquake generation is similar throughout the region. It certainly does not require that the system is in a state of self organized criticality.

Figure 1 shows the result of the calculation of D_2. Successive curves are generated by taking the data in successive groups of 3, 6, 9 ... and so on so that each curve corresponds to an increase of the dimension of the embedding space by three. The observed dimensionality of the attractor should also increase until the dimensionality of the embedding space is large enough for the attractor. Once there is enough room in the embedding space the dimension will then arrive at a constant value m, the dimension of the attractor. This dimension measured from the slopes of the curves in Figure1 is between 5 and 6 for main shocks above magnitude 4. For mainshocks of magnitude above 2 the constant slope of the curves may be between 11 and 12. For the catalog containing all the events of magnitude above 2, including aftershocks the slope of the curves becomes constantbetween 3 and 4.

Several things can cause the inconsistency of the results from the calculations for the catalogs which contain mainshocks whose magnitudes are above 4 and above 2. Errors can be introduced into the calculation due to incomplete recording of small events. Since the data used are time intervals between successive events, missing events could make the data appear more noisy and lead to an apparent higher dimensionality. More noise can also be brought into the mainshock catalogs by the operation of removing aftershocks from the catalog. In addition to the fact that the aftershock windows are strictly empirical the errors introduced into the main shock catalog become significant for the small mainshock events whose epicenters generally have a large uncertainty. This affects the selection of the space window for removing aftershocks.

The low dimensionality shown in the catalog which contains all the events indicates that the aftershocks are an integral part of the whole dynamics and contain significant information about the dynamics of the lithosphere. Removing aftershocks from the catalog eliminates this useful information about the lithosphere dynamics. It may also be that aftershock events are characterized by a smaller length scale in the attractor than that of mainshocks since they are local events of mainshocks both in time and space. This means that the section of the attractor corresponding to aftershocks has a lower dimension than that of mainshocks and this reduces the overall effect of the dimension across the attractor to less than what we obtained from the mainshock catalog.

The main source of error probably comes from missing events in the catalog. Since the lithosphere is a spatially open system, when we use a space window to select a study area, some events which are significant to the dynamics may be left outside of the window. It is not clear however how those missing events will affect the reconstruction of the phase portrait. Another source of error is due to the way of investigating the scaling of the correlation function. Ruelle (1990) showed that the correlation dimension calculated from a data series of length N is meaningless unless the correlation dimension is less than or equal to $2\log_a N$. He argued that the measurement should be taken at least over one decade, therefore, a should be 10. Only those dimensions which are well below $2\log_{10} N$ have credibility. In the above exercise only the analysis for the catalog containing all the events meets this requirement. The analysis for the catalog that contains mainshocks of magnitudes above 4 is far from it. The region within which the correlation function behaves linearly in the log-log plot is only across a scale of 3 for the mainshock catalog with magnitude above 4.

APPLICATION OF SINGULAR SPECTRUM ANALYSIS TO EARTHQUAKE CATALOG

The method proposed by Grassberger and Procaccia (1983) for computing the correlation dimension of a strange attractor is powerful, but its validity for a short and noisy data series such as an earthquake catalog is questionable (Ruelle, 1990). Nevertheless, with such short and noisy records there may be some useful integral parameters other than the correlation dimension; they may be better behaved and still indicate some of the underlying order in the data. Singular spectrum analysis, is an application of the Karhunen-Loeve expansion for random processes (Fukunaga, 1972) and shows promise of being able to characterize chaotic data.. The applications of the method to the paleoclimatic time series, which contain only a few hundred data samples, and to the earthquake catalog from the Parkfield, California showed significant results (Vautard and Ghil, 1989; Horowitz, 1989).

Singular spectrum analysis (SSA) is discussed in detail by Broomhead and King (1986). We note here that like the analysis which leads to the D_2 measure it also begins with a segmentation of a time series into delayed signals each of which we take to have a length of 20 samples. The delayed signals can be arranged into a trajectory matrix and this matrix can be analysed by the well known methods of singular value decomposition. The eigen values that result from this decomposition will in an ideal case consist of a few with "large" values and the rest which merge into a "noise" floor. The number of large eigen values is a measure of the number of basis vectors required to approximate the coherent part of the signal. The coherence integral method on the other hand can be thought of as reacting to the way points are clustered in neighborhoods in the embedding space. Singular spectrum analysis yields dimensionality bounded by integers only.

The spectra shown in figure 2 result from applying SSA with an n-window of 20. The spectra for the catalogs containing mainshocks of magnitude above 4 and 2 and the catalog containing all the events of magnitude above 2 do not show the simple structure proposed by Broomhead and King in which the singular values decrease monotonically until they merge into the flat noise floor. Several plateaus occur in each spectrum. Surprisingly in all three cases, there are three singular values above the first plateau which is probably associated with a noise floor. This strongly supports the result from the correlation dimension analysis for the catalog containing all the events in the last section. It supports the argument that aftershocks contain a significant amount of information about the integral dynamics of the lithosphere.

DISCUSSION AND CONCLUSION

Both the analysis of the correlation dimension and the singular spectrum analysis suggest the existence of an attractor of dimension

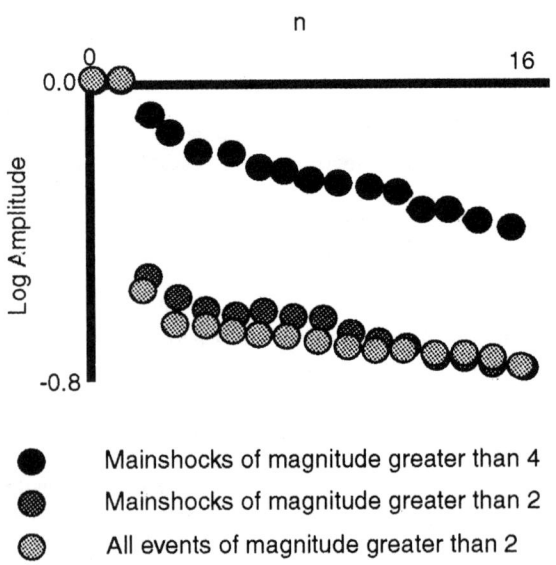

Fig. 2. The results of the SSA as derived by Li (1991).

between 3 and 4 for our study area near Vancouver, Canada. Such analysis is for a specific region; if the similarity of seismicity world wide (Keilis-Borok, 1990) is indeed valid, this result may be generalized. Since the catalogs used in the analysis have marginal homogeneity, the final conclusion can be drawn only after the analysis has been done for more areas of the world.

The fact that low dimensionality is associated with a catalog dominated by aftershocks may mean only that the attractor is the process associated with aftershock generation. This is interesting in itself but an additional speculation is warranted in this area. The study area is in the northern part of the Cascadia subduction zone and turbidites in cores from off-shore Washington and Oregon suggest that occasional very large earthquakes may have occurred in this area prior to the arrival of European scientists on this coast (Adams 1990). The last such event may have occurred a few hundred years ago and it is tempting to speculate that the low dimensionality of the attractor in the small magnitude seismicity for the area reflects aftershock activity from this event. If it is evidence of fossil aftershock activity it might also reflect the 1964 Alaska earthquake.

Acknowledgments. Comments and suggestions by V. I. Keilis-Borok were very inspiring and helpful. Discussions with Andrei Gabrielov and Leon Knopoff also helped us shape these arguments. We thank D.H. Weichert for his work on the earthquake catalog of the west coast of Canada and Bosko Loncarevic for reading and commenting on the manuscript.

REFERENCES

Adams, John, Palaeoseismicity of the Cascadia Subduction Zone: Evidence from Turbidites off the Oregon Washington Margin, *Tectonics*, 9, 4, pp569-583, 1990.

Broomhead, D. S. and G. P. King, Extracting qualitative dynamics from experimental data. *Physica D*, 20, 217-236, 1986.

Brown, D., Q. Li, E. Nyland, and D.H. Weichert, Premonitory seismicity patterns near Vancouver Island, Canada. *Tectonophysics*, 167, 299-312, 1989.

Burridge, R. and L. Knopoff, Model and theoretical seismicity, *Bulletin of the Seismological Society of America*, 57, 342-371, 1967.

Carlson, J. M. and J. S. Langer, Mechanical model of an earthquake fault, *Physical Review A*, 40, 6470-6484, 1989.

Dieterich, J. H., Time-dependent friction and the mechanics of stich-slip, *Pure and Applied Geophysics*, 116, 790-806, 1978.

Dieterich, J. H., Modelling of rock friction 1. Experimental results and constitutive equations, *Journal of Geophysical Research*, 84, 2161-2168, 1979.

Froehling, H., J. P. Crutchfield, J. D. Farmer, N. H. Packard and R. S. Shaw, On determining the dimension of chaotic flows, *Physica D*, 3, 605-617, 1981.

Fukunaga, K., *Introduction to Statistical Pattern Recognition*, Academic Press, New York, 1972.

Gardner, J. K. and L. Knopoff, Is the sequence of earthquakes in southern California, with aftershocks removed Poissonian?, *Bulletin of the Seismological Society of America*, 64, 1363-1367, 1974

Grassberger, P. and I. Procaccia, Measuring the strangeness of a strange attractor, *Physica D*, 9, 189-208, 1983a.

Griggs, D. T. and D. W. Baker, The origin of deep focus earthquakes, in *Properties of Matter under Unusual Conditions*, Mark & Fernback, Interscience, New York, 1969.

Henon, M., Numerical exploration of Hamiltonian systems, in *Chaotic Behavior of Deterministic Systems*, eds. L Gerard, H. G. Robert and S. Raymond, North-Holland Publishing Company, Amsterdam, 1983.

Horowitz, F., A strange attractor underlying Parkfield seismicity, *EOS*, 70, 1359, 1989.

Huang, J. and D. L. Turcotte, Are earthquakes an example of deterministic chaos?, *Geophysical Research Letters*, 17, 223-226, 1990.

Kagan Y. Y. and L. Knopoff, Random stress and earthquake statistics: Time dependence, *Geophysical Journal of the Royal Astronomical Society*, 88, 723-731, 1987.

Kagan, Y. Y., Spatial distribution of earthquakes: The 3-point moment function, *Geophysical Journal of the Royal Astronomical Society*, 67, 697-717, 1981a.

Kagan, Y. Y., Spatial distribution of earthquakes: The four-point moment function, *Geophysical Journal of the Royal Astronomical Society*, 67, 719-733, 1981b.

Keilis-Borok, V. I., The lithosphere of the earth as a nonlinear system with implications for earthquake prediction, *Reviews of Geophysics*, 28, 19-34, 1990.

Keilis-Borok, V. I., L. Knopoff, I. M. Rotwain and C. R. Allen, Intermediate-term prediction of occurrence times of strong earthquakes, *Nature*, 335, 690-694, 1988.

Keilis-Borok, V. I., V. M. Podgayetskoya and A. G. Prozoroff, On the local statistics of catalogs of earthquakes, *Computational Seismology*, 5, 55-79, 1971.

Li, Q., Patterns in Chaos; Nonlinear Geodynamics of Earthquake Prediction, *Ph.D. Thesis*, University of Alberta, Edmonton, 1991.

Lorenz, E. N., Deterministic non-periodic flow. *Journal of Atmospheric Sciences*, 20, 130-141, 1963.

Ogawa, M.., Shear Instability in a ViscoElastic Material as the Cause of Deep Focus Earthquakes, *Journal of Geophysical Research*, 92, 13801-13810, 1987.

Packard, N. H., J. P. Crutchfield, J. D. Farmer and R. S. Shaw, Geometry from a time series, *Physical Review Letters*, 45, 712-716, 1980.

Roux, J. C., R. H. Simoyi and H. L. Swinny, Observation of a strange attractor, *Physica D*, 8, 257-266, 1983.

Rudnicki, J. W., Physical models of earthquake instability and precursory processes, *Pure and Applied Geophysics*, 126, 531-554, 1988.

Ruelle, D., Deterministic chaos: the science and the fiction, *Proceedings of the Royal Society of London* A., 427, 241-248, 1990.

Ruina, A. L., Slip instability and state variable friction laws, *Journal of Geophysical Research*, 88, 10359-10370, 1983.

Takens, F., Detecting strange attractors in turbulence, in *Lecture Notes in Mathematics, 898*, Springer, Heidelberg-New York, 1981.

Vautard, R. and M. Ghil, Singular spectrum analysis in nonlinear dynamics, with applications to paleoclimatic time series, *Physica D*, 35, 395-424, 1989.

Q. Li, 6268 Dalmarnock Crescent, Calgary N.W., Alberta, Canada
E. Nyland, Department of Physics, University of Alberta, Edmonton, Alberta, Canada, T6G 2J1

Dynamics of a Seismic Regime: Vrancea - A Case History

CEZAR-IOAN TRIFU[1] AND MIRCEA RADULIAN

Center of Earth Physics and Seismology, Bucharest, Romania

Detailed analysis of the intermediate depth ($h > 60$ km) seismic regime, based on a microearthquake catalog extended over 18 years (3870 events), confirms the existence of two active zones in Vrancea, between 60-110 km (A) and 120-170 km depth (B). Non-linear features such as distinct jumps in the frequency-magnitude distribution and spatial clusterings are clearly emphasized when restricting to an active zone only. A geometrical similarity is noticed between them, and a scaling factor of 2.5 is found (B/A). According to a discrete source model, it is equivalent to a scaling in fragmentation, and explains a shift of 0.6 magnitude units observed between the frequency-magnitude distribution of events inside each zone. Two parameters, the ratio of small to moderate events and the fractal dimension of depth distribution show significant variations of about 500% and 20%, respectively, prior to a strong earthquake in zone B (30 Aug 1986). No variation can be associated with strong eathquakes in zone A (4 Mar 1977 and 30 May 1990).

INTRODUCTION

Recent studies concerning the dynamics of the seismic regime in Vrancea region (Romania) emphasized a number of features drawing attention to the non-linearities of the earthquake process in a large scale-length, from microearthquakes to major events [*Trifu and Radulian*, 1989; 1991]. Thus, non-linear behaviors were revealed in both the size and spatial distributions of earthquakes. Meanwhile, non-linear time evolution of parameters related to these distributions was associated with the occurrence of major events, and consequently considered as defining precursory effects [*Radulian and Trifu*, 1991]. It was concluded that the non-linear behavior became extremely apparent when the analysis was restricted to a distinct seismic area of the fault, denoted as active zone. Two active zones were pointed out in Vrancea between 60-110 km (A) and 120-170 km depth (B).

In their previously referenced work, the authors introduced a discrete source model (DSM) as an attempt to explain a large amount of observational data in Vrancea including the above mentioned non-linearities. DSM prescribes a discrete structure of the active zone: there is a minimum, elementary area for the earthquake generation (S_o), and an elementary area for the strength distribution (S_a) along the active zone (asperity-cells). As suggested by data, these two areas are roughly equal ($S_o \simeq S_a$). There are three ways to release seismic energy on the fault: the failure of a low-strength elementary surface through a crack-like mechanism, the failure of a high strength asperity-cell or group of linked asperity-cells through an asperity-like mechanism, and the failure of the whole active zone through percolation.

The present paper proposes a comprehensive analysis of the Vrancea intermediate depth seismic regime in time, space and energy domains, based on updated observations, for the last 18 years (1974-1991). Nonlinearities of the earthquake process are emphasized by a comparative study of the active zones. These results are further interpreted in terms of the DSM, reviewing the main features of this model, and making an attempt to include the differences between the active zones. This region hosted 3 major events within the above time interval. Consequently, the subject of earthquake predictability is finally discussed, based on the non-linear behavior of some parameters describing the seismic regime, such as the ratio of small versus moderate events, and the fractal dimension of space distribution.

---

[1] Now at Department of Geological Sciences, Queen's University, Kingston, Ontario.

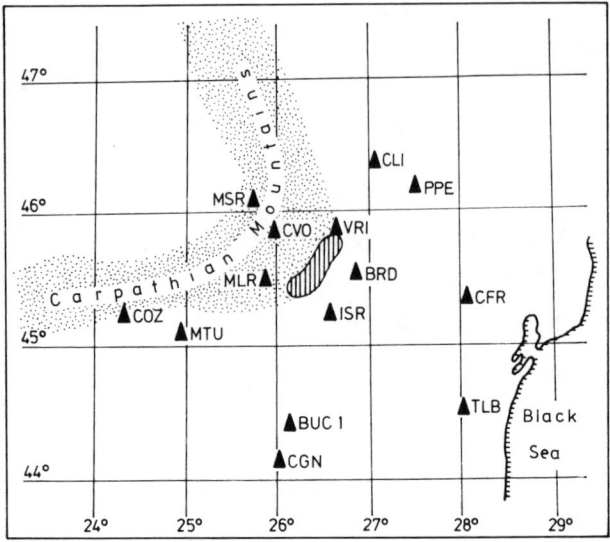

Fig. 1. Local seismic telemetered network; the hatched surface is the epicentral area of Vrancea intermediate depth earthquakes ($h > 60$ km).

OBSERVATIONS

The Vrancea region in Romania has a well localized seismic activity mainly consisting of intermediate depth earthquakes ($h = 60\text{-}180$ km) in an extremely confined epicentral area of about 50 by 20 km. Since 1980, the seismic survey of this region has been made by a radio telemetered 14-station local network of vertical component short-period Teledyne-Geotech S13 velocity transducers (Figure 1). Each station has two electronic magnification channels, and a central digital acquisition unit is operating in real time. Precise earthquake locations emphasize the morphology of the seismic volume (Figure 2). The fault plane is striking NE-SW and dips nearly vertical towards NW. The scarcity of seismicity between 40 and 60 km depth can be noted.

To study the seismic regime over a broad magnitude range, it is worth considering events recorded at less than 3 stations. *Trifu et al.* [1990] and *Trifu and Radulian* [1991] proposed a method for depth and magnitude determination from recordings at one or two stations only, consisting of calibrations with time difference between P and S arrivals (t_{S-P}) and total signal duration (τ): $h = f(t_{S-P})$, $M_L = g(\tau, t_{S-P})$. To this purpose, accurate locations ($\sigma_h \leq 4$ km) of almost 300 events by a joint hypocenter technique (JHD) were used, based on digital readings at a minimum of 7 stations. By choosing the VRI and MLR stations, which are situated in the epicentral area and characterized by highest magnifications, a homogeneous and completeness tested ($M_L \geq 2.6$) Vrancea intermediate depth microearthquake catalog for the time interval October 1980 - December 1991 was set up. The local magnitude scale in use is moment calibrated.

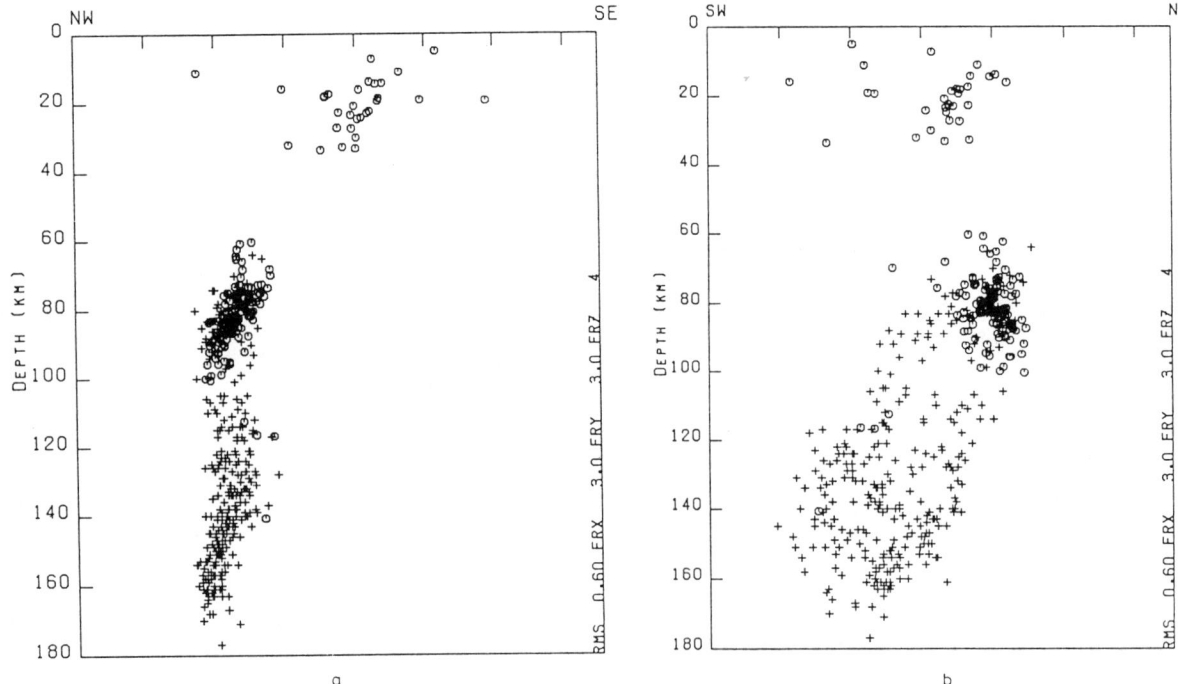

Fig. 2. NW-SE (a) and SW-NE (b) vertical projections of hypocentral distributions; high accurate locations ($\sigma_x = \sigma_y \leq 3$ km, $\sigma_h \leq 4$ km) are represented by crosses for events occurring between 1981-1987, and circles for those occurring within 40 days following the major earthquake of 30 May 1990; horizontal scales are similar to the vertical ones.

Because the VRI and MLR stations have been both in operation since 1974, we are currently extending our catalog to include the time interval January 1974 - September 1980. The identification of intermediate depth earthquakes and the t_{S-P} and τ readings has been done on vertical component recordings of short period SKM3 displacement transducers. The t_{S-P} reading error is 0.1 s and τ is measured until signal/noise $\simeq 2$. To test any possible bias between our readings on the two types of instruments (S13 and SKM3), simultaneous readings over a 7-month time interval have been compared. The data shown in Figure 3 follow the distributions

$$t_{S-P}^{S13} = 0.98 t_{S-P}^{SKM} + 0.36$$
$$\tau^{S13} = 0.98 \tau^{SKM} + 2.48 \quad (1)$$

A good correlation is obtained, which consequently allows us to extrapolate the above calibration to the SKM3 data. The completeness of the catalog was tested for the day-to-night cultural noise by the sum phasor technique of *Rydelek and Sacks* [1989], as shown in Figure 4, and magnitude thresholds $M_L = 2.2$ and 2.6 were obtained. However, these completeness thresholds are somewhat higher on SKM3 ($M_L = 2.6$ and 3.0) than on the S13, due to the much lower sensitivity of the displacement transducer at high frequencies, which are characteristic of small events.

The earthquake catalog contains a number of 3870 events ($M_L \geq 1.5$) occurring over 18 years (1974-1991). It includes 3 major events, 4 March 1977, $h = 93$ km, $M_w = 7.5$; 30 August 1986, $h = 131$ km, $M_w = 7.3$, and 30 May 1990, $h = 90$ km, $M_w = 6.9$ (Figure 5). Associated aftershock intervals of abnormal activity of 1, 2 and 5 months, respectively (Figure 6), were removed before analysing the earthquake depth distribution over the whole time interval (Figure 7). *Trifu and Radulian* [1991] related the two-lobe shape of this distribution to the existence of two seismic zones A and B, with distinct activities centered around 70-110 km and 120-160 km depth, respectively. Their hypothesis is also supported by other arguments, such as the depth distribution of the moment tensor principal axes, and stress inhomogeneity, and will be adopted hereafter.

ANALYSIS OF NON-LINEAR CHARACTERISTICS OF THE SEISMIC PROCESS

Time Distribution of Seismic Activity

The earthquake time series was analysed to determine its clustering properties. To begin with, observed time distributions were compared with a Poisson distribution corresponding to a random process. The probability of k events occurring in a prescribed time interval Δt is

$$P(k; \lambda \Delta t) = \frac{(\lambda \Delta t)^k}{k!} \exp(-\lambda \Delta t) \quad (2)$$

Here, the λ parameter is a constant related to density of events on the time axis. The larger λ is, the higher the probability $1 - P(0; \lambda \Delta t) = exp(-\lambda \Delta t)$ of finding at least one event in Δt is. If T is the catalog time duration and N the total event number, then

$$\lambda \Delta t \simeq T/N \quad (3)$$

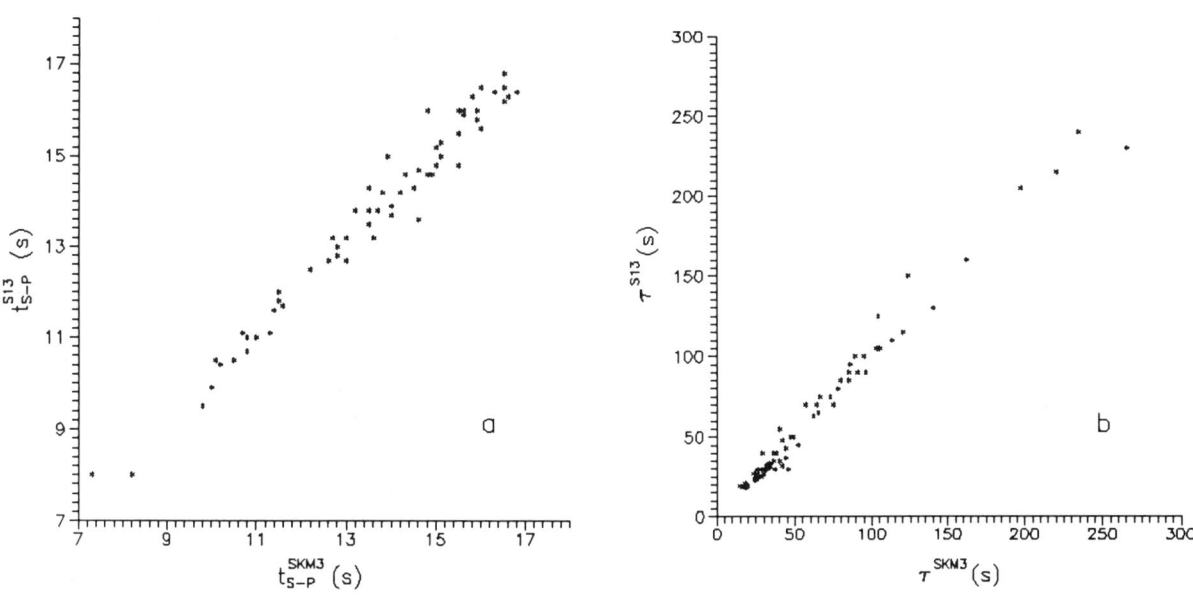

Fig. 3. Comparison between analogue readings from velocity (S13) and displacement (SKM3) transducers: (a) time differences between S and P wave arrivals; (b) total durations.

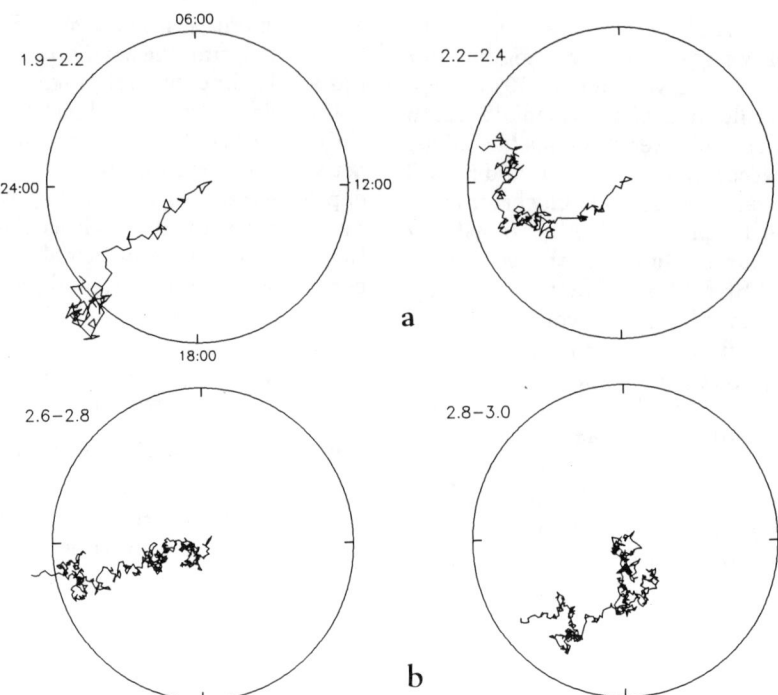

Fig. 4. Phasor sum plots for testing the magnitude completeness of the microearthquake catalog on different depth intervals: (a) $h = 60\text{-}115$ km, (b) $h > 115$ km; the numbers in each frame represent the magnitude interval; the catalog is complete to a 95% confidence level when the phasor sum remains within the circle.

Since the tested data sets (for different time and depth intervals) fit the Poisson distribution well (Figure 8), the occurrence of the Vrancea intermediate depth earthquakes can be considered as a pure random process, to a first approximation.

A refined analysis of the clustering in time can be provided by fractal statistics [*Smalley et al.*, 1987]. The time of occurrence is a point on the time axis. The total duration of the sequence is divided into smaller time intervals, $\tau = T/n, n = 2, 3, 4,...$, and for each case the number of intervals which contain at least one event (x) is determined as a function of interval length (τ). If a power-law dependence $x = x(\tau)$ is obtained, then the time series has a fractal behavior. In other words, a scalelength invariant time clustering appears. The corresponding fractal dimension D can be computed from [*Mandelbrot*, 1982]

$$x \sim \tau^{1-D} \quad (4)$$

D has subunitary values; the clustering is higher as D approaches 0, while a value of 1 corresponds to a uniform distribution (events equally spaced in time).

The number of occupied intervals x as a function of τ is plotted in Figure 9, together with the corresponding uniform distribution. The smallest τ interval is 2 hours, while the other time intervals are taken as powers of 2. The fractal dimension D is estimated from the linear part of the observed curves: 0.87, 0.84, and 0.89, for cases a, b, and c, respectively. The mean value of

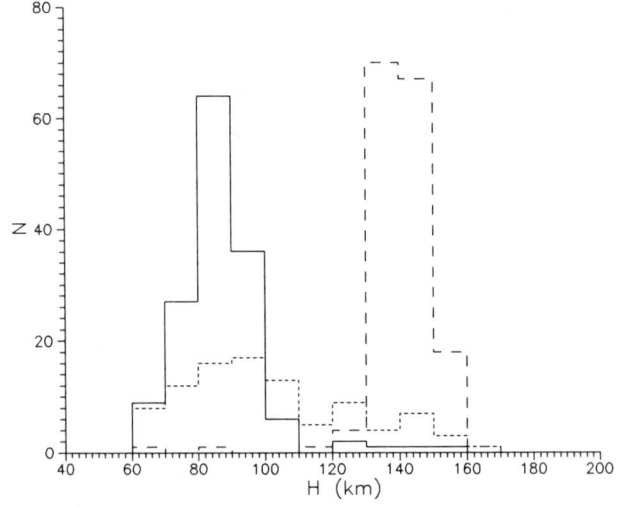

Fig. 5. Depth distribution histograms for one month of aftershock activity following the major events of 4 Mar 1977 (dotted line), 30 Aug 1986 (dashed line), and 30 May 1990 (solid line).

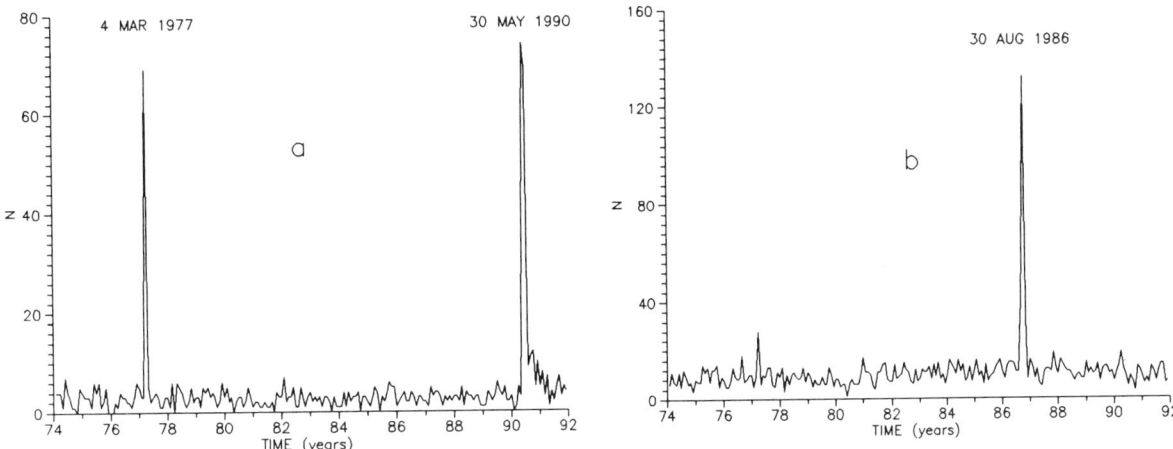

Fig. 6. Monthly seismic activity (1974-1991, $M_L \geq 2.6$): (a) $h = 60\text{-}115$ km (zone A); (b) $h > 115$ km (zone B).

0.87 may be due to a slight tendency towards clustering, but this is not relevant due to the curvature of the observed distribution (a continuous curvature measures the randomness of time occurrences).

Fractal Analysis of Earthquake Space Distribution

To study the seismicity clustering along the slab, the spatial analysis will be restrained to depth, the only coordinate accurately determined for all events. Due to the confined epicentral area, the depth dependence is likely to be the most important parameter.

The depth distribution has a fractal structure if

$$C(r) \sim r^D \tag{5}$$

where D is the fractal dimension, and $C(r)$ is the correlation integral [*Grassberger and Procaccia*, 1983]

$$C(r) = \frac{2}{N(N+1)} K_{R<r} \tag{6}$$

In the above equation, N is the total number of events in a given subset, and $K_{R<r}$ is the number of earthquake pairs having the distance R between their depths smaller than r.

The time variation of D is studied by a moving window technique applied to the whole time series, taking subsets of 150 events at a 10 event shift interval. For each subset, D is computed by least squares as the slope of the log-log graph of (5) for r ranging from 5 to 17 km, with a 3 km step. Figure 10 shows the results for the two active zones, together with the corresponding standard deviations (σ_D).

Three anomalies of $15-20\%$ are found in the deeper zone only, $h > 115$ km (Figure 10b). *Radulian and Trifu* [1991] associated them with foci clusterings around certain depths. They made a comparative analysis on smaller depth intervals and found that the first two anomalies successively occurred around 150 to 160 km, and 140 to 150 km. The third anomaly is given by the aftershock activity following the 1986 major event, and corresponds to its focal region, mainly restrained to 130 to 140 km. These clusterings are coalescing effects due to changes in the average distances between events, and do not necessarily concern increases in the earthquake number in a certain depth range. This can be seen by comparing Figure 10b with Figure 5b.

The upper active zone does not show any significant anomaly, although it hosted two major earthquakes in

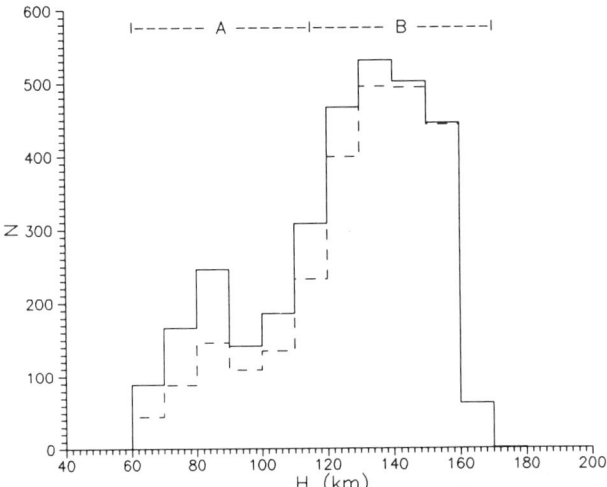

Fig. 7. Depth distribution histograms (1974-1991) for $M_L \geq 1.5$ (solid line), and $M_L \geq 2.6$ (dashed line); aftershock intervals of 1, 2, and 5 months, associated with the strong earthquakes of 1977, 1986, and 1990 respectively, have been removed.

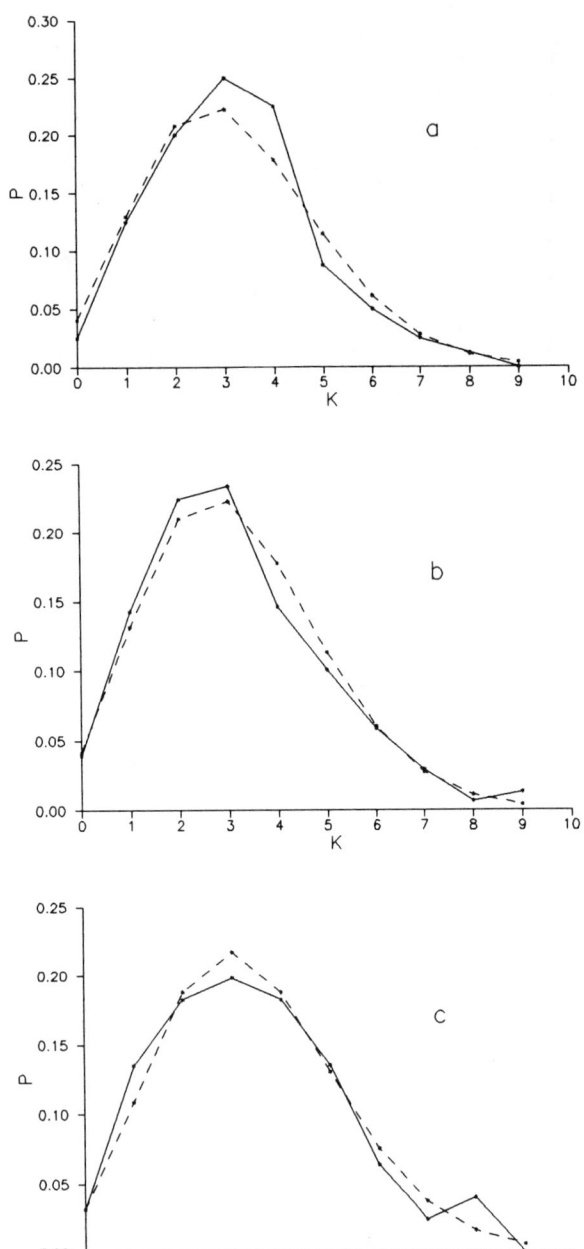

Fig. 8. Probability of finding k events with $M_L \geq 3$ in a given Δt time interval; observations (solid lines) are tested with Poisson distributions (dashed lines): (a) 1 Apr 1977 - 29 May 1990, $h = 60\text{-}115$ km, $\Delta t = 60$ days; (b) 1 Jan 1974 - 29 Aug 1986, and (c) 1 Nov 1986 - 31 Dec 1991, both having $h > 115$ km, $\Delta t = 15$ days.

in a subset. However, since this explanation fails for the 1990 aftershock activity, for which we have a large number of earthquakes, the existence of a different physical mechanism is likely to be involved.

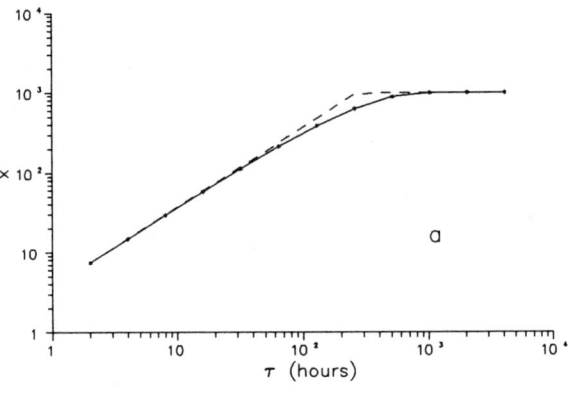

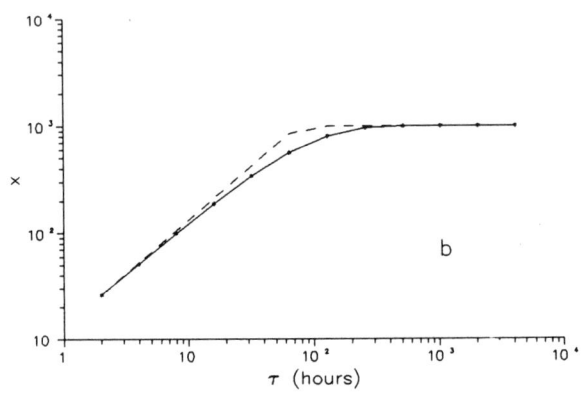

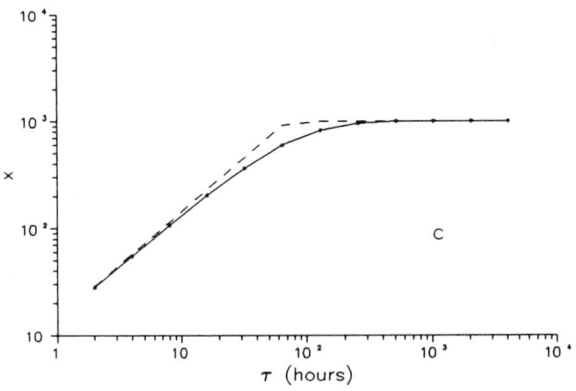

Fig. 9. Distribution of the fraction x of occupied time intervals, for $M_L \geq 2.6$, as a function of time interval τ; the fractal dimension is given by the deviation of the observed distribution (solid line) from the corresponding uniform one (dashed line); (a) 1 Apr 1977 - 29 May 1990, $h = 60\text{-}115$ km; (b) 1 Jan 1974 - 29 Aug 1986, and (c) 1 Nov 1986 - 31 Dec 1991, both having $h > 115$ km.

1977 and 1990. A possible explanation for this strikingly different behavior could be the much lower resolution in this case as a consequence of a lower seismicity level. This limitation is imposed by the method used, which requires a statistically sufficient number of events

Earthquake Size Distribution

The frequency-magnitude distribution in the Vrancea region emphasizes non-linear features at both the scale of the seismic cycle (Figure 10), and a smaller scale (Figure 11). In the former case, a strong deficit in the number of earthquakes is noticed between moderate and large magnitudes ($M_S \simeq 5.5-6.5$) as compared with a linear dependence $\log N = f(M)$, known as the Gutenberg-Richter (GR) law. This behavior was also observed in several other areas of the world, as discussed by *Trifu and Radulian* [1989]. In the latter case, an intermediate relative enhancement is pointed out, which is interpreted in the frame of the DSM as a magnitude threshold for asperity-like earthquakes.

Since the above non-linearities in the size distribution are clearly scale effects, they are easily delineated for an appropriate space-time scale. Indeed, Figure 12 shows the relevance of restricting the analysis of the frequency-magnitude distribution to a seismically active zone. This zone is taken as a particular area of the fault, able to generate both a characteristic seismicity and a characteristic major event. The moving window technique points to a clear jump in this distribution for Vrancea when passing from one active zone to another. As discussed in the second section, these results are statistically significant when $M_L \geq 2.6$ for zone A and $M_L \geq 3.0$ for zone B. Shape similarity inside each zone emphasizes the main non-linear trends of

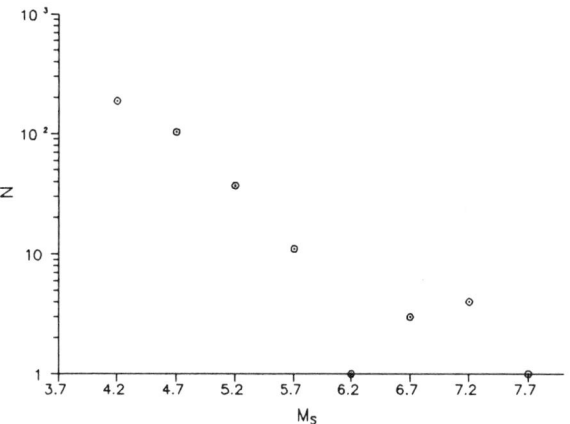

Fig. 11. Noncumulative frequency-magnitude distribution of Vrancea intermediate depth earthquakes with $M_S \geq 4$ occurred between 1936 and 1991; N values are computed for intervals of 0.5 magnitude units centered on M_S.

the distributions in Figure 11. When the statistical randomness tests [*Trifu and Radulian*, 1991] are also taken into account, the intermediate bumps around $M_L = 3.3$ for zone A, and $M_L = 3.9$ for zone B, are shown to be real.

According to the DSM, there are two possibilities that can explain the shift in magnitude between the two zones:

(i) If the fragmentation at the earthquake scale is invariant, the average number of asperity-cells per unit surface is constant, independent of the active zone. Thus, $S_o{}^A = S_o{}^B$, where $S_o{}^A$ and $S_o{}^B$ are typical asperity-cell surfaces for the two zones. Since the inhomogeneity parameter $\varepsilon = S_o/S$, where S is the area of the whole source, is also constant [*Trifu*, 1987], $S^A = S^B$ for the asperity-like events of minimum magnitude. The relationship determined by *Kanamori and Anderson* [1975] allows us to estimate the seismic moment

$$M_o = \Delta\sigma S^{3/2} \quad (7)$$

It follows that a difference in magnitude between the two zones (ΔM_L) leads to a corresponding difference in stress drop. For $\Delta M_L \simeq 0.6$ (Figure 11) we estimate this difference by using (7) and the log moment-magnitude relation for Vrancea intermediate depth earthquakes [*Trifu and Radulian*, 1991]

$$\log M_o = 1.0 M_L + 17.4 \quad (8)$$

and obtain

$$\Delta\sigma^B/\Delta\sigma^A = 10^{\Delta M_L} \simeq 4 \quad (9)$$

Since such an important difference is not observed in the

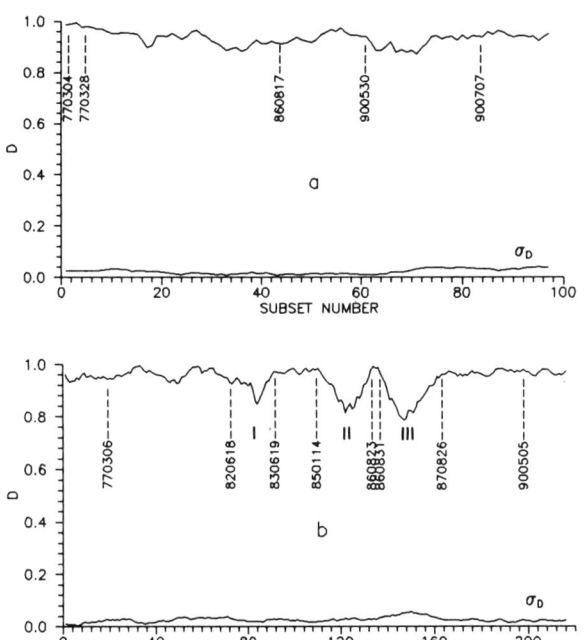

Fig. 10. Temporal variation of the spatial fractal dimension D: (a) $h = 60-115$ km, $M_L \geq 2.2$; (b) $h > 115$ km, $M_L \geq 2.6$; the standard deviation is shown at the bottom of each frame; anomalies are denoted by I, II and III.

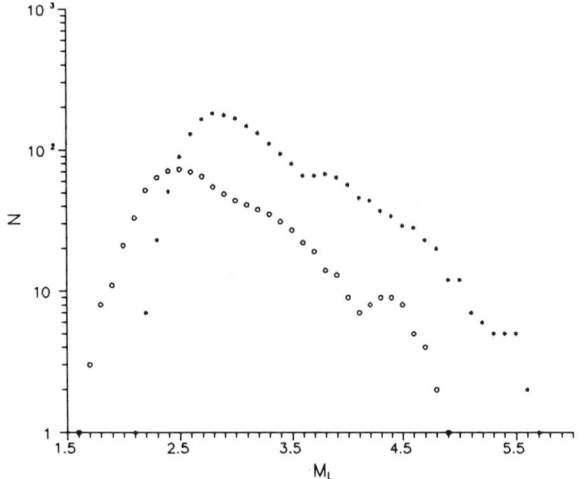

Fig. 12. Noncumulative frequency-magnitude distributions for zone A (open circles) and zone B (solid circles); events between 1974 and 1991 are considered after removing the aftershock intervals (see also Figure 5).

data [*Oncescu*, 1986], the hypothesis of fragmentation invariance must be rejected for the time being.

(ii) If the stress drop is assumed constant, the observed ΔM_L shift is due to a variation in fragmentation between the two active zones, according to (7). Since ε is constant,

$$S_o^B/S_o^A = S^B/S^A = 10^{2/3 \Delta M_L} \simeq 2.5 \quad (10)$$

For a given active zone, the surface S_o characterizes both the asperity-cell (the elementary strength nucleus of an asperity-like failure), and the minimum surface which does not lock after the rupture (the elementary shear stress-free surface). We relate this surface to an instability mechanism for the seismic process in each zone, governed by specific state conditions. It is likely that the sharp change in fragmentation of the lithospheric material around 115 km depth is the consequence of a phase transition it undertakes.

Another aspect which deserves attention when comparing distributions in Figures 12 and 13 is the shift between the cut-off magnitudes of zones A and B ($M_L = 4.9$ and 5.7, respectively). DSM assumes that the major earthquake is generated by a percolation proces. When a critical fraction of stress-free surface is developed over the fault, the inception of a major event becomes possible. The critical density depends entirely on geometrical parameters: the total active zone area and the elementary fault surface. As can be evaluated from the hypocentral distributions in Figure 2, the total active zone areas are roughly in the same ratio as the elementary surfaces, and it follows that a scaling factor $\nu = 2.5$ could be considered between the two zones. Consequently, the shift in cut-off magni-

tudes equals that in threshold magnitudes of asperity-like earthquakes, $\Delta M_L = 0.6$, in good agreement with the data. Note the scaling is also present in the aftershock frequency-magnitude distribution, and a similar magnitude shift between the two zones is observed (Figure 14).

A POSSIBLE DRIVING MECHANISM FOR EARTHQUAKES

The detailed study of the Vrancea seismic regime revealed a series of non-linearities which must be accounted for as constraints in the modeling of the earthquake process. At the tectonic scale, there are strong variations of the seismic activity along the slab (Figures 2, 6, and 7): a relative moderate seismicity of shallow earthquakes ($h = 0-40$ km, $M_S \sim 5$), a quiescence below the crust ($h = 40-60$ km), and two intermediate depth active zones able to generate major earthquakes ($M_w \sim 7$), sharply separated by a transition zone ($h = 110-120$ km). The active zones are separate, distinct segments, resulting from the presence of large scale heterogeneities along the fault plane. Each of them has its own parameters governing seismic activity at the scale of the earthquake cycle, their interaction being neglected to a first approximation.

In our view, the active zone has a discrete structure of a characteristic dimension resulting from fragmentation of the lithospheric material under specific state conditions. We assume the elementary surfaces (cells) of the fault may have three levels of shear strength during the cycle: high for asperity-cells, low for unruptured cells, and zero for ruptured cells (shear stress-free surfaces). Meanwhile, three characteristic earthquake mechanisms acting at different hierarchical levels are proposed: crack-like failure for the smallest magnitude events (background seismicity); asperity-like fail-

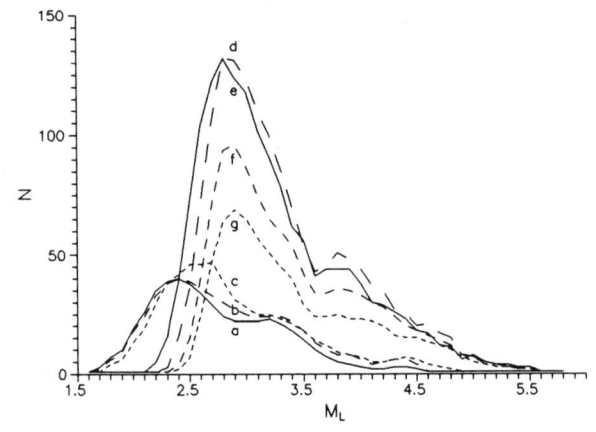

Fig. 13. Noncumulative frequency-magnitude distributions on depth intervals; zone A: (*a*) 60-90 km; (*b*) 70-100 km; (*c*) 80-110 km; zone B: (*d*) 120-150 km; (*e*) 130-160 km; (*f*) 140-170 km; (*g*) > 145 km.

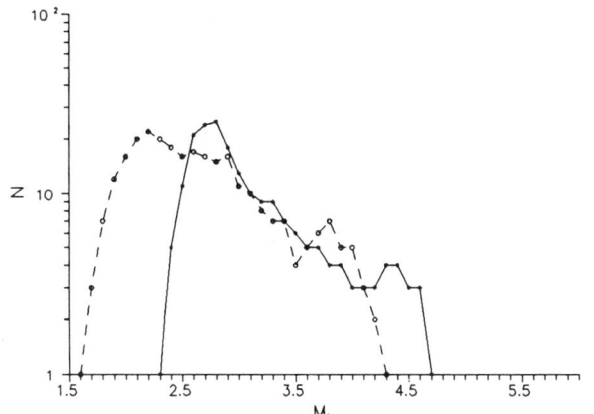

Fig. 14. Noncumulative frequency-magnitude distributions for one-month time interval of aftershocks: 1986 (solid line) and 1990 (dashed line).

ure for larger events, and percolation for the major earthquakes. The presence of different mechanisms generates different earthquake families, which are detected in the frequency-magnitude distributions as significant deviations from self-similarity (GR). Imposing specific constraints, such as a non-linear pattern, significantly reduces the number of possible model classes that might explain the generation of the earthquake phenomena [Lomnitz-Adler et al., 1991]. Our model takes into consideration the stress transfer only in a geometrical sense. Note that Ben-Zion and Rice [1993] recently arrived to similar non-linear GR statistics allowing the stress transfer along asperity and non-asperity regions to follow the elastic dislocation theory.

The background seismicity and the asperity distribution over the fault are responsible for both the coupling of the tectonic stress along the slab and the time evolution of the seismic activity on the active zone. The background seismicity continuously generates a shear stress-free surface on the fault, creating the necessary condition for the inception of asperity-like earthquakes. An asperity failure is possible only if it is surrounded by a sufficient level of shear stress-free surface. Later on, when this surface reaches a critical value, a major cluster suddenly develops by the percolation of the active zone, making possible the triggering of the major event. Thus, percolation explains the magnitude gap between moderate and large earthquakes in the frequency-magnitude distribution (Figure 11). In effect, it is equivalent to a diffusion process of the weak surface over the fault.

Radulian et al. [1991] numerically simulated the earthquake generation process during a cycle for the Vrancea lower intermediate depth active zone, by assuming a series of time-invariant parameters, such as the active zone total area, elementary cell area, background seismicity rate, and asperity-like earthquake magnitude threshold. Through a comparative analysis, the results of the present study emphasize the dependence of the above parameters on each active zone. This variation cannot be attributed only to a difference in fragmentation of the lithospheric material. Indeed, the average ratio of the background seismicity rates in zones B and A, except for the aftershock intervals, is around 5. Since the fragmentation acts only as a 2.5 scaling factor, a supplementary factor of 2 remains to be considered by the DSM in the generation of the background seismicity into zone B. It is still by far to be justified. However, note that unlike the spatial scaling, a temporal scaling has no meaning, as a tremendous deviation from the above ratio is obtained when comparing aftershock sequences (Figures 5 and 6).

EARTHQUAKE PREDICTABILITY: CHANCE OR CERTAINTY?

The γ parameter expresses a relative variation of small versus moderate events in a given active zone. It is computed as

$$\gamma = N_1/(N_2 + s) \qquad (11)$$

where N_1 and N_2 are the number of small and moderate earthquakes, and s is a smoothing factor accounting for the scarcity of data. The earthquake magnitude domains chosen for the computation of N_1 and N_2 values in zone B are $M_L = 3.3 - 3.9$ and $M_L > 4.2$, and a shift of 0.6 magnitude units was considered for zone A. γ is determined by using large enough time steps to reduce the statistical fluctuations, $\Delta t = 5$ months and 10 months, respectively. The factor of 2 between the active zones restrains the time resolution in zone A. The temporal variation of the γ parameter is comparatively represented in Figure 15. A sharp anomaly, 5 times the noise level, precedes the major event of 1986 in zone B by about a year, while there is no significant perturbation in zone A associated with the major events occurring there. Since the statistical fluctuations lead to errors of the order of $N^{1/2}$, relative errors of about 15% can be estimated for γ.

The time evolution of the fractal dimension of the depth distribution D was already presented in a previous section (Figure 10). Note, the precursory anomalies in zone B can be interpreted as clusterings of shear stress-free surfaces bringing the major asperity into relief [Trifu, 1990]. The lack of any significant precursory variation of all the above parameters within zone A is certainly a striking feature of this analysis. We consider that the somewhat lower time resolution in this zone cannot entirely hide a potential effect at a reduced time scale. For example, the different behavior of the D parameter for aftershocks of the 1977 and 1990 earthquakes is obviously not explained by a different resolution, since rich statistics are available for them.

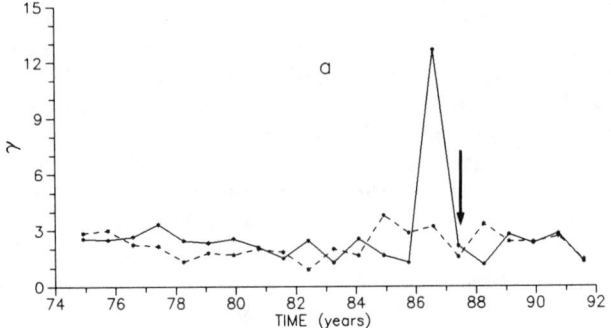

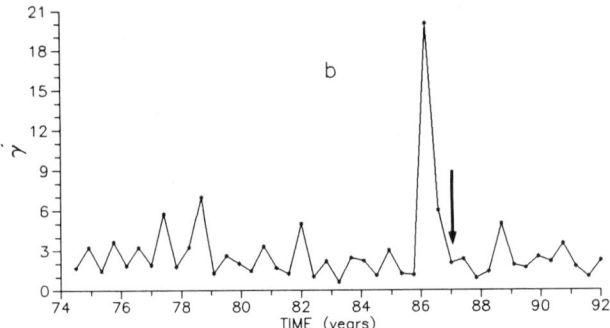

Fig. 15. Temporal variation of the γ parameter for zone A (dashed line; $s = 2$) and zone B (solid line; $s = 1$); $(a)\Delta t = 10$ months; $(b)\Delta t = 5$ months

Basically, it is the presence of the active zones, implying characteristic earthquakes, which gives a chance for long term prediction. The analysis of the seismic regime pointed out significant differences between the Vrancea active zones. The lack of repeatability raises a serious doubt concerning the validity of the above parameters as predictive descriptors, but the correlation of the differences in the seismic regime of each active zone with the possibility to generate a major earthquake must be further investigated to clarify this matter.

CONCLUSIONS

The present study of the Vrancea seismic regime based on a homogeneous and complete microearthquake catalog extended over 18 years confirms the existence of two active zones, lying at depths of 60 to 110 km (zone A), and 120 to 170 km (zone B), respectively. A geometrical similarity is observed between them which is quantitatively measured by a scaling factor of 2.5 (B/A). In the case of a constant stress drop in the two zones, it is equivalent according to DSM with a scaling in fragmentation of the lithospheric material, and explains the observed shift effect of 0.6 magnitude units in the corresponding number-size distributions of events.

The limitation of the analysis to an active zone allows the best delineation of the non-linear characteristics of the seismic process. They appear in energy and space, as jumps in the frequency-magnitude distribution, associated with different types of earthquake failure (crack-like, asperity-like, percolation), and spatial clusterings creating shear stress-free surfaces around major asperities, respectively.

Time domain analysis reveals both the constancy of the seismic regime (time clustering is neglectable), and the randomness of earthquake inter-occurrences. A factor of 2 between the average seismicity rates in the two zones cannot be explained simply by a time scaling, since this factor strongly changes when comparing aftershock sequences. Consequently, the time behavior of one zone with respect to the other reflects not only geometrical, but also physical differences in the shear stress balance on the fault.

Two parameters, namely the ratio of weak to moderate events, and the fractal dimension of depth occurrences seem to offer precursory information in case of the strong earthquake of 30 August 1986 (zone B). However, nothing similar is emphasized in case of other strong events: 4 March 1977 and 30 May 1990 (zone A). Taking into account the error levels provided by appropriate analysis, we conclude that the variation of these parameters within zone B reflected real effects. Meanwhile, lower statistics within zone A cannot be uniquely invoked to explain the lack of repeatability in precursory variations. This consequently suggests that other important specific differences in the behavior of the active zones exist, and further investigations are required to point them out.

Acknowledgments. We thank D. Jansen for helping us in preparing the final version of this paper.

REFERENCES

Ben-Zion, Y., and J.R. Rice, Earthquake failure sequences along a cellular fault zone in a 3D elastic solid containing asperity and non-asperity regions, *J. Geophys. Research*, in print, 1993.

Grassberger, P., and I. Procaccia, Measuring the strangeness of strange attractors, *Physica D*, 9, 189-208, 1983.

Kanamori, H., and D.L. Anderson, Theoretical basis of some empirical relations in seismology, *Bull. Seismol. Soc. Am.*, 65, 1073-1095, 1975.

Lomnitz-Adler, J., D. Comte, and M. Pardo, Models of seismic fracture: constraints imposed by the magnitude-frequency relation (abstract), paper presented at the XXth General Assembly, IUGG, Vienna, Aug. 11-24, 1991.

Mandelbrot, B.B., *The Fractal Geometry of Nature*, W.H. Freeman, New York, 1982.

Oncescu, M.C., Some source and medium properties of the Vrancea seismic region, Romania, *Tectonophysics*, 126, 245-258, 1986.

Radulian, M., and C-I. Trifu, Would it have been possible to predict the August 30, 1986 Vrancea earthquake?, *Bull. Seismol. Soc. Am.*, 81, 2498-2503, 1990.

Radulian, M., C-I. Trifu, and F.O. Cărbunar, Numerical simulation of the earthquake generation process, *Pure Appl. Geophys.*, 136, 499-514, 1991.

Rydelek, P.A., and I.S. Sacks, Testing the completeness of earthquake catalogues and the hypothesis of self-similarity, *Nature*, 337, 251-253, 1989.

Smalley, R.F., J-L. Chatelin, D.L. Turcotte, and R. Prévot, A fractal approch to the clustering of earthquakes, *Bull. Seism. Soc. Am.*, 77, 1368-1381, 1987.

Trifu, C-I., Depth distribution of local stress inhomogeneities in Vrancea Region, Romania, *J. Geophys. Res.*, 92, 13878-13886, 1987.

Trifu, C-I., Detailed configuration of intermediate seismicity in Vrancea region, *Rev. Geofis.*, 46, 33-40, 1990.

Trifu, C-I., and M. Radulian, Asperity distribution and percolation as fundamentals of an earthquake cycle, *Phys. Earth Planet. Inter.*, 58, 277-288, 1989.

Trifu, C-I., and M. Radulian, Frequency-magnitude distribution of earthquakes in Vrancea: relevance for a discrete model, *J. Geophys. Res.*, 96, 4301-4311, 1991.

Trifu, C-I., M. Radulian, and E. Popescu, Characteristics of intermediate depth microseismicity in Vrancea region, *Rev. Geofis.*, 46, 75-82, 1990.

M. Radulian, Center of Earth Physics and seismology, . P.O.Box MG-2, Bucharest-Măgurele, Romania.

C-I. Trifu, Department of Geological Sciences, Queen's University, Kingston, Ontario K7L 3N6, Canada.

The Precursor of Instability for Nonlinear Systems and Its Application to Earthquake Prediction
---the Load-Unload Response Ratio Theory

XIANG-CHU YIN[1], CAN YIN AND XUE-ZHONG CHEN

Institute of Geophysics, State Seismological Bureau, Beijing, P.R.China

By analysing the unstabilizing process of nonlinear systems, we found that the instability precursor is significant rise of the response rate or response ratio. Applying this idea to earthquake prediction, we take the periodically varying stress in lithosphere caused by tide-generating force for loading and unloading and choose the crust deformation, gravity, seismicity etc., as the responses. If the ratio of the response rate during loading period to that during unloading period can be measured, it will quantitatively predict the degree of danger due to the imminent earthquake in a given region.

With the data of ten large earthquakes($Ms \geq 7.0$) that occurred in the Chinese Mainland during 1970-1992, the Loma Prieta Earthquake ($Ms=7.1$, Oct. 18,1989), Landers Earthquake ($Ms=7.5$, June 28,1992), etc., it is found that the response ratios of most strong earthquakes increase distinctly before the main shocks. After that, using this method, we successfully predicted several earthquakes including the Datong Earthquake($Ms=5.8$, March 26,1991) and the Puer Earthquake ($Ms=6.3$, Jan. 27,1993)

INTRODUCTION

Earthquake prediction is an extremely difficult problem. Many scientists have made great efforts and got some inspiring achievements in this field, but in more cases they failed. From their grievous experiences, most seismologists have recognized that the fundamental way to solve this problem should lie in the transition from empirical prediction to that based on physics (Ding Guo-yu et al., 1979; Fu Cheng-yi, 1989; Geller, R. J., 1991; T. Rikitake, 1976; Yin Xiang-chu, 1987).

What is the physical essence of an earthquake? It is just the instability of the focal medium, accompanied by rapid release of strain energy. This is the essential distinction between earthquakes and other tectonic movements (such as orogenesis, tectonic movement of plate and creep of faults etc.) (Yin Xiang-chu,1991; Yin Xiang-chu and Yin Can, 1991). Since the lithosphere is a highly nonlinear system (Keilis-Borok, 1990) in order to predict earthquakes effectively, we have to analyse the unstabilizing process of nonlinear systems.

In some of our previous papers (Yin Xiang-chu, 1987,1991,1993; Yin Xiang-chu and Yin Can,1991), the principal idea of response ratio theory has been expounded systematically and analysis of various nonlinear systems has been carried out. For convenience of reading, we'll take material test for an example and analyse its result briefly.

The trait of any linear systems consists in constant response rate $\Delta R/\Delta P$ (output parameter / input parameter, or state parameter / controlling parameter, or response / loading, we adopt the last one in this paper). Shown in Fig.1 is that if test material is ideally linear elastic, then the relationship between load P (P here refers to stress σ) and response R (R is strain ϵ here) will be linear (solid line in Fig. 1(b)) and the response rate will be constant (in this case, response rate is the reciprocal of Young's modulus of the tested material). For real materials, however, the relation between R and P can not remain linear when stress exceeds the elastic limit of material (dotted line in Fig. 1(b)).

For convenience, we modified Fig. 1(b) into Fig. 2. It is seen that once the deformation of material exceeds the elastic phase, the system will become non-linear. From then on, its response rate will no longer be constant but get more and more great as the system approaches instability. When it reaches the instability point (the summit T in Fig. 1(b)), the response rate becomes infinite.

$$\lim_{\Delta P \to 0} \frac{\Delta R}{\Delta P} = \infty \qquad (1)$$

This is just the definition of instability of a system.
Increasing of response rate as the system tends to be unstable is an elementary precursor, and other precursors would be

---

[1]Also at Graduate School of Acedemia Sinica, Department of Earth Sciences, Beijing,PRC.

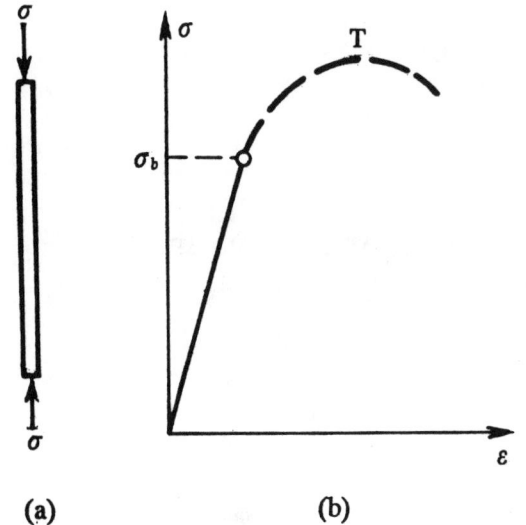

Fig.1. Uniaxial compression test

deduced from it, such as the intensifying of fluctuation (that is, a tiny disturbance ΔP may cause great change in state of system, and the intensification of fluctuation would further trigger instability in turn), etc.. People have recognized such kind of phenomena in daily life long ago. "It is the last straw that breaks the camel's back". This proverb vividly depicts such a phenomenon. In nonlinear science, the so called "butterfly effect" also describes the same thing with precise mathematic method which means that a very tiny variation in boundary condition would induce dramatic change for the solution in large scale for a nonlinear system (if a butterfly quivers its wings in Brazil, that could cause a tornado in Texas!)

Since response rate $\Delta R/\Delta P$ is a dimensional parameter, its value will depend on the unit adopted. It is convenient to introduce the non-dimensional response ratio Y_0. Suppose the response rate is $\Delta R_0/\Delta P_0$ when system is stable (Fig. 2), and it becomes $\Delta R_1/\Delta P_1$ when load becomes P_1, we can define the first type of response ratio as

$$Y_0 = \frac{\Delta R_1/\Delta P_1}{\Delta R_0/\Delta P_0} \qquad (2)$$

If a system is stable, $\Delta R_1/\Delta P_1$ ought to be equal to $\Delta R_0/\Delta P_0$ and $Y_0 = 1$; while state deviates from stability, $Y_0 > 1$; and when a system becomes unstable, $Y_0 \to \infty$. Thus, it can be seen that the response ratio can be used to measure quantitatively the degree of instability of a system.

However, Under most circumstances, it is difficult to determine $\Delta R_0/\Delta P_0$, so we turn to the second type of response ratio Y. When a nonlinear system is close to instability, the response rate to loading significantly differs from that to unloading. So we define the second type of response ratio as:

$$Y = \frac{\Delta R_+/\Delta P_+}{\Delta R_-/\Delta P_-} \qquad (3)$$

where $\Delta R_+/\Delta P_+$ is response rate to loading while $\Delta R_-/\Delta P_-$ refers to unloading. For this reason, Y is also called load-unload response ratio. It can quantitatively indicate the degree of the imminency of instability for a nonlinear system as well as Y_0 and can be measured more conveniently.

APPLICATION OF RESPONSE RATIO THEORY TO EARTHQUAKE PREDICTION

Now let's come back to seismology. If we could load and unload the lithosphere medium of some region again and again to measure ΔR_+ and ΔR_- of the lithosphere corresponding to loading and unloading respectively, we would work out its response ratio Y and then determine quantitatively the degree of instability in this region or the imminence of earthquake.

How to load a block of crust of over thousands cubic kilometers? Certainly it can't come true by manpower at present. It is nature that gives us such means, that is the earth tide. Tide-generating force changes periodically for ever, hence the stress in crust induced by tide also varies periodically. Provided having observed the response ΔR_+ and ΔR_- to load and unload in this crust block, we would calculate Y and judge the perilous degree of a strong earthquake by the value of Y.

For the purpose of earthquake prediction, the crust deformation, tilt, level of groundwater, seismicity and other seimic parameters, along with geoelectric or geomagnetic parameters etc. could be choosen as responses.

PRACTICAL EXAMPLES

We've analysed all earthquakes (Ms $\geq$ 7.0) that occurred in the Chinese Mainland during 1970-1992, of which the three earthquakes of Qinghai, Tonghai and Yijitaicuo are omitted because of the scant data before mainshocks, so there are data for ten earthquakes still available. For each strong earthquake, we delimit its examined region (taking the epicenter as the center of the region with linear dimension 1°-2°) and then arrange the smaller earthquakes which occurred in this region before the main earthquake in time order (no limit on their magnitudes). We take N earthquakes as one group (say, N=100) and calculate the effective shear stress increment $\Delta \tau_e$ caused by tide on the fault

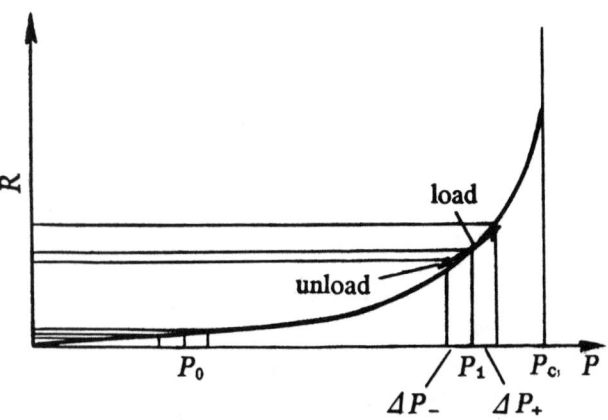

Fig.2 Response vs. general load for a nonlinear system

plane at the moment each quake occurred. Then for every group we judge each earthquake whether it lies in duration of which $\Delta\tau_e$ is positive or it lies in duration of which $\Delta\tau_e$ is negative. Afterwards, add up mth power of energy E_i of every earthquake lying in positive durations to make $(\Sigma E_i^m)_+$. The same process is used to get $(\Sigma E_i^m)_-$. The proportion of $(\Sigma E_i^m)_+$ to $(\Sigma E_i^m)_-$ is used for load-unload response ratio Y

$$Y = \frac{(\sum_{i=1}^{n_+} E_i^m)_+}{(\sum_{i=1}^{n_-} E_i^m)_-} \quad (4)$$

Where the power m can be selected among following values: 0, 1/3, 1/2, 2/3 and 1. When $m=1/3$, E^m indicates the linear scale of the earthquake focus; when $m=2/3$, E^m expresses the area scale; while $m=1/2$, E^m is called Benioff strain. It is proved in practice that the effect when $m=1/2$ or $1/3$ is the most suitable.

Now, we will briefly explain how to calculate $\Delta\tau_e$. We take Dziewonski's Preliminary Reference Earth Model. The elastic deformation of the earth in this model is described by six first order differential equations (P.Melchior, 1987). Following the work of Molodensky and Takeuchi (P.Melchior, 1987), we can calculate each of the stress components caused by tide-generating force on any arbitrary cross-section in any point of the crust using numerical integration method developed by Runge-Kutta. We needn't give the detailed algorithm here since it has been solved long before.

Supposing on a selected fault plane, tide-generating force causes shear stress and normal stress (taking tensile stress for positive), then the effective shear stress caused by tide on this plane will be

$$\overline{\tau}_e = \overline{\tau}_n + f \cdot \sigma_n \cdot \frac{\overline{\tau}_n}{|\overline{\tau}_n|} \quad (5)$$

where f is friction coefficient, and the increment of effective shear stress is defined as follows

$$\Delta\tau_e = \overline{\tau}_e \cdot \frac{\overline{D}}{|\overline{D}|} \quad (6)$$

where D is the slip vector on earthquake fault plane.

For a certain region (to be more precise, a point in the earth lithosphere), we can calculate $\Delta\tau_e(t)$ caused by tide there. We take the average fault parameters, which comes from (Zhang Cheng et al., 1990; Zhang Zhao-cheng et al., 1990) and other references, for the fault parameters of every earthquakes because

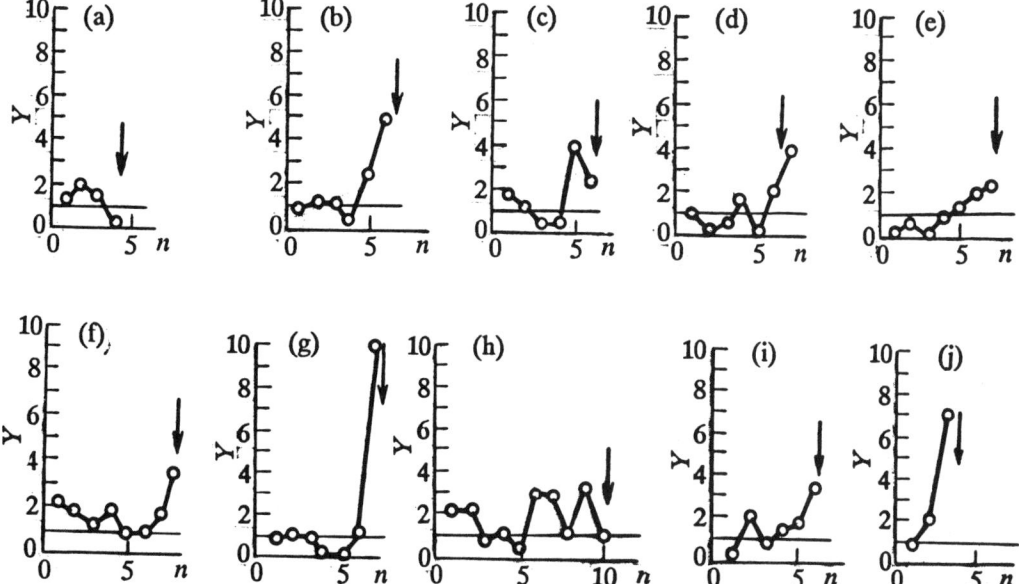

Fig.3 Load-unload response ratio vs. relative time before the ten strong earthquakes (Ms7.0) occurred in Chinese Mianland during 1970-1992. (a) 1973,12,06 Luhuo earthquake (Sichuan Province) Ms=7.6; (b) 1974,05,11 Yongshan earthquake (Sichuan Province) Ms=7.1; (c) 1974,08,11 Wucha earthquake (Xinjiang Autonomous Region) Ms=7.3; (d) 1975,02,04 Haicheng earthquake (Liaoning Province) Ms=7.3; (e) 1976,05,29 Longling earthquake (Yunnan Province) Ms=7.4; (f) 1976,07,28 Tangshan earthquake (Hebei Province) Ms=7.8; (g) 1976,08,16 Songpan earthquake (Sichuan Province) Ms=7.2; (h) 1985,08,23 Wucha earthquake (Sichuan Province) Ms=7.1; (i) 1988,11,06 Lancang earthquake (Yunnan Province) Ms=7.6; (j) 1990,04,26 Gonghe earthquake (Qinghai Province) Ms=7.0

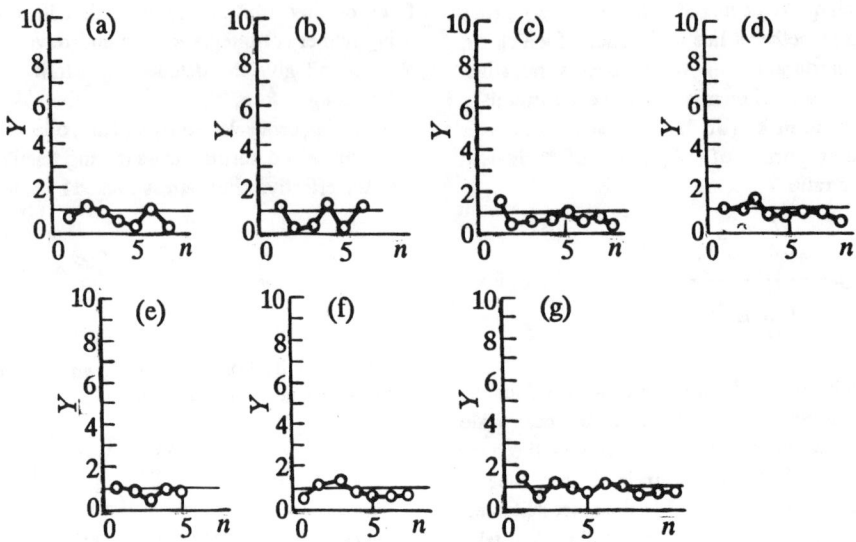

Fig.4 Load-unload response ratio vs. relative time in seven regions with stable crust during 1970-1992
(a) Southern Tanlu Fault (35.5°N±1°,118°E±1°); (b) Northern Shaanxi Province (40.5°N±1°,109°E±1°); (c) Eastern Sichuan Province (31°N±1°,105°E±1°); (d) Northern Shandong Province (37°N±1°,119°E±1°); (e) Western Shandong Province (37°N±1°,117°E±1°); (f) Northern Henan Prorvince (35°N±1°, 113°E±1°); (g) Southern Shandong Province (35°N±1°,117°E±1°)

it is hard to get the focal mechanism of each small earthquake. The time period for $\Delta\tau_e(t) > 0$ is called positive duration while that for $\Delta\tau_e(t) < 0$ is called negative duration. In positive duration, the effective shear stress caused by tide-generating force on the seismic fault strengthens original tectonic effective shear stress and would be advantageous to the occurrence of earthquake. On the contrary, $\Delta\tau_e$ in negative duration suppresses the occurrence of earthquake. Since $\Delta\tau_e$ is several orders smaller than the ordinary tectonic stress, under normal condition the effect of $\Delta\tau_e$ must be negligible so that $(\Sigma E_i^m)_+$ is roughly equal to $(\Sigma E_i^m)_-$, thus $Y \approx 1$. But when the selected region is unstable (on the eve of a strong earthquake), any tiny disturbance will cause tremendous response and make the value of Y very large.

In the ten earthquakes that we have examined, there are eight or nine earthquakes for which Y value rises distinctly before main shock (Fig.3). Besides, we chose seven areas with relatively stable crust (no large earthquakes these years and plentiful data records. Then we calculated Y during 1970-1992 and found that Y always fluctuates slightly around 1 in more than two decades (Fig. 4). Therefore, these results strongly indicate that the value of Y contains abundant seismogenic information indeed.

In each chart from Fig. 3 to Fig.6, the ordinate is load-unload response ratio defined in formula (4); The abscissa refers to time or relative time. We take several months as the length of time window to calculate the load-unload response ratio Y (Fig.5,

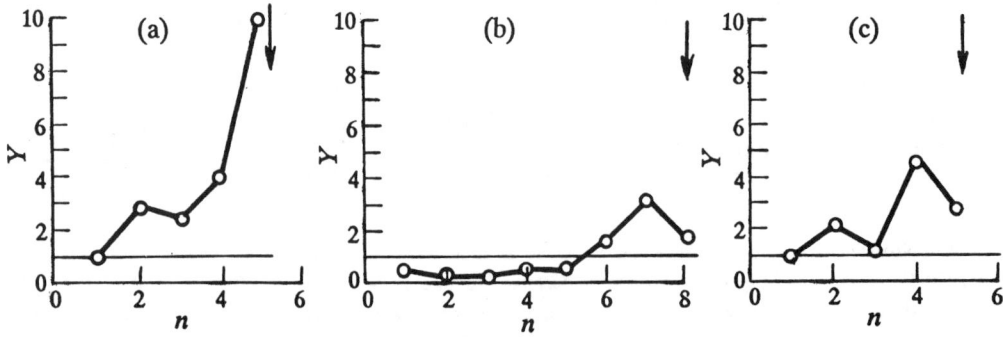

Fig.5 Load-unload response ratio vs. time for three famous earthquakes in recent years.
(a) 1989,10,18 Loma Prieta Earthquake (Ms=7.1); (b) 1992,06,28 Landers Earthquake (Ms=7.5);
(c) 1988,12,07 Spitak Earthquake (Ms=6.9)

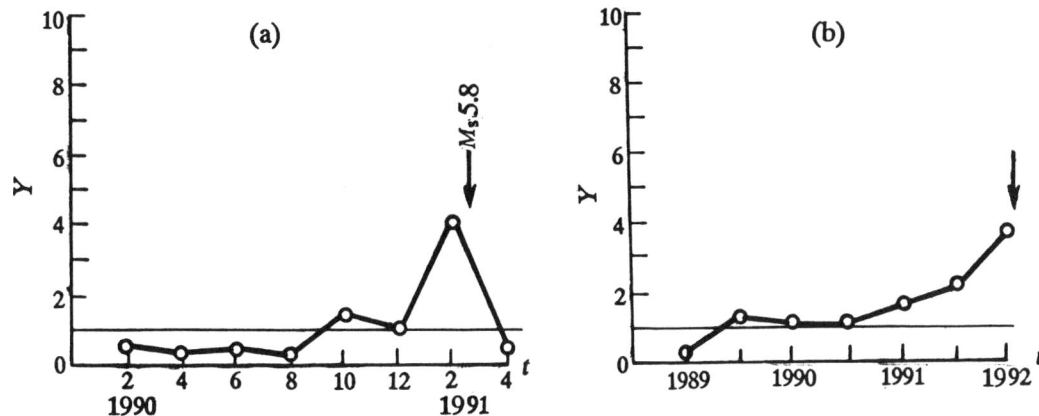

Fig.6 Y-t charts before some predicted succesfully earthquakes. (a) 1990,03,26 Datong Earthquake (Shanxi Province) Ms=5.8; (b) 1993,01,27 Puer Earthquake (Yunnan Province) Ms=6.3

Fig.6). Sometimes, there are only very few even no events in some time windows for the examied region due to the heterogeneous seismicity. For this resson we take N earthquakes (N ranges from 50 to 100) as a group to calculate a value of Y and the abscissa denotes relative time (the tactic order of groups) in such case (Fig.3, Fig.4).

In addition, some earthquake cases outside China have been examined in this way, including Loma Prieta Earthquake (California, USA, $Ms=7.1$, Oct.18,1989), Landers Earthquake (California, USA, $Ms=7.5$, June 28,1992) and Spitak Earthquake (Armenia, CIS, $Ms=6.9$, Dec.7,1988). It can be seen from Fig.5 that the Y value increase significantly much more than 1 before all these events. The time window length of abscissa are (a) 1 year; (b) 1 year; and (c) 7 months respectively. The duration during which the value Y is greater than 1 ranges from 2 to 4 years.

After the retrospective examination described above, we have successfully made some practical predictions (mid-term) in real time. Figure 6 shows the value Y vs. time before (a) Datong Earthquake (Shanxi Province, Eastern China , $Ms=5.8$, March. 26, 1991) and (b) Puer Earthquake (Yunan Province, Southwestern China, $Ms=6.3$, Jan.27,1993). According to the pronounced consecutive increasing of Y, we predicted successfully them on March 21, 1991 and Jan.8,1993 respectively.

The data concerning Chinese earthquakes in this paper was offered by the Center of Analysis and Prediction, State Seismological Bureau of P.R.China, which is the authoritative institution of China in this field. The catalog of California came from the Seismic Network of Univ. of California, Berkeley and the catalog of Spitak earthquake from the Institute of Physics of the Earth, Academy of Science of Russia. Since Y is the ratio of all responses in positive period to that in negative period, the effect of systematic error due to magnitude of earthquakes is rather small.

It must be noted that the above idea of predicting earthquake by means of response ratio is related to, but distinct from the idea to correlate earth tide with seismicity. In the former case, seismicity, which is characterized quantitatively with earthquake energy or its m-th power, is adopted only as the response of a specific block of crust to the tiny variation of crust stress state in order to estimate the instability degree of the region. Besides seismicity, a lot of other geophysical parameters can be chosen as response also, such as crust information, gravity, tilt, ground water level and seismic wave parameters etc. In fact, several Chinese excellent geophysists have been worked on this topic. Their results will be published in the near future. In the latter one, the relation between earth tide and seismicity has been being a controversial topic, for which quite different opinions appeared in the literature (Knopoff,1964; Klein,1976; Heaton,1975,1982; Kilston and Knopoff,1983; Varga,1985).

Acknowledgements. The authors are grateful to Professor Cheng-yi Fu, Gong-xu Gu, Xin-ling Qin, Shi-rong Mei, A. Sobelev, S. L. Yunga and Dr. E.A.Bergman et al. for their earnest help and support. This subject is supported by Natural Science Foundation of China under grant 49070185 and the Chinese Joint Seismological Science Foundation under grant JSSF90053.

REFERENCES

Ding Guo-yu, Ma Tsung-chin and Mei Shi-rong, Methods of earthquake prediction, in *Earthquake Prediction*, pp453-466, TERRAPUB, Tokyo Unesco, Paris, 1979.

Fu Cheng-yi, *Geophysics in China in the eighties*, pp1-10, Academic Press, Beijing, 1989 (in Chinese).

Geller, R.J., Shake-up for earthquake prediction, *Nature, 352*, pp275-276,1991.

Heaton.T.H., Tidal triggering of earthquakes, *Geophys.J.R.astr. Soc.,43*,307-326,1975.

Heaton.T.H., Tidal triggering of earthquakes, *Bull. Seismol. Soc. Am., 72*, 2181-2200, 1982.

Kilston.S. and L.Knopoff, Lunar-solar periodicities of large earthquakes in southern California, *Nature,304*,7-8,1983.

Klein.F.W., Earthquake swarms and the semidiurnal solid earth tide, *Geophys.J.R.astr.Soc.,45*,245-295,1976.

Knopoff. L., Earth tides as a triggering mechanism for earthquakes, *Bull.Seismol.Soc.Am.,54*,1865-1870,1964.

Varga.P., Influence of external forces on the triggering of earthquakes, *Earthq. Predict. Res. 1*, 191-201, 1985.

Keilis-Borok, The lithosphere of the Earth as a nonlinear system with implications for earthquake prediction, *Review of Geophysics,28*,19-24,1990.

Melchior, P., *The Tides of the Planet Earth*. Pergamon Press, 1978.

Rikitake,T., *Earthquake Prediction*, Elsevier, Amsterdam, 1976.

Yin Xiang-chu, Exploring the new approach of earthquake prediction, *Earthquake Research in China,3*, 1-7,1987.

Yin Xiang-chu, A new paremeter measuring the instability degree of the crust for a specific region—the ratio of loading and unloading response and its application to geological disaster forecast. *Proceedings of the PRC/USSR Workshop on Geodynamics and Seismic Risk*, Beijing, China,1991.

Yin Xiang-chu, A new appoach to earthquake prediction, *PRERODA* (Russia's "Nature"), No.1, 21-27,1993 (In Russian).

Yin Xiang-chu and Yin Can, The precursor of instability for nonlinear systems and its application to earthquake prediction, *Science in China (B), 34, 8*, 977-986,1991.

Zhang Cheng et al., *Focal Mechanism in China*, Academic Press, Beijing, 1990 (in Chinese).

Zhang Zhao-cheng et al., *Earthquake Cases in China*, Vol.1-Vol.4, Seismological Press, Beijing,1990 (in Chinese).

Xiang-chu Yin, Can Yin and Xue-zhong Chen, Institute of Geophysics,SSB,Beijing 100081,P.R.China.

Strange Attractor in Nonlinear Fluctuations of Length of the Day (LOD) Time Series

R.K. TIWARI, J.G. NEGI, AND K.N.N. RAO

National Geophysical Research Institute, Hyderabad 500 007, India

The evidence for "low dimensionality" and "chaos" in length of the day (LOD) time series is of considerable interest. The variations in LOD time series on subannual time scales are a proxy for global angular momentum, which bears a direct relation on the dynamics of the coupled atmosphere-ocean system. The axial rotation of the Earth, as represented by (LOD) time series, has been examined using the time delay embedding theory and (G-P) Algorithm. The analysis reveals a low fractional dimension of the chaotic attractor of the order of 5 to 7. The indication of low dimensionality suggests that system can be adequately described by at least 5 to 7 independent key variables. The result is confirmed even after filtering dominant terms of Earth's tidal spectrum. The Kolmogorov entropy analysis (< 0.13/day) gives an intrinsic time scale of the order of 7-8 days. The relevance of the results is also discussed in connection with the coupling of annual and subannual atmospheric-ocean interaction processes.

INTRODUCTION

The study of the causes of diverse variations in the rotation of the Earth is one of the most intriguing and fundamental concern of classical mechanics and geophysics. The Earth's rotation is not strictly circular and its fluctuations are of the order of a few milliseconds per century. The development of modern space-based geodetic techniques, and a vast amount of accurate data on Earth's rotation, provides complex information. The researchers and scholars of solid Earth geophysics, oceanography, paleontology, meteorology and astronomy are discovering various Earth-Moon-Sun relationships and complexity of atmospheric, oceanic and convection in the Earth's interior.

Our planet Earth continuously changes orientation in space under the influence of external torques generated by gravitational coupling with the Sun-Moon system and the other in the solar system and also internal torques resulting from the dynamic geophysical processes involving redistribution of masses and angular momentum [David, 1989]. The term Earth orientation can be categorized as: (i) the angular speed (Z-component or changes in the rate of rotation and the length of the day LOD), and (ii) the polar motion (X, Y components). The axial component (Z-component) is the magnitude of the angular velocity. The polar motion ("equatorial" or X, Y components) arises from the displacement of the instantaneous rotation axis from the Earth's figure axis (as well as from a fixed direction in space) [Chao, 1984]. The rotation of the Earth exhibits minute but complicated changes measurable up to several parts in 10^8 in speed to correspond to a variation of several milliseconds in Earth axial rotation and even larger variations in polar motion. The fluctuation in the LOD is actually measured in terms of (UT_1–AT). Here, UT_1 is universal time determined by astrometric/geodetic means with respect to a constant "AT" (86400 seconds) kept by an atomic clock [Chao, 1984]. It is directly proportional to the perturbation in the axial component of the angular momentum. In the present analysis we are concerned with the variations in Length of the Day (LOD) as the time derivative of UT_1–AT.

The dynamic Earth, during the course of its rotational and revolutional journey, interacts with a wide variety of geophysical and astronomical disturbances perturbing and slowing Earth's rotational speed. These changes occur over a broad spectrum of time scales, ranging from days to centuries and longer [Dickey et al., 1987]. Accordingly, these variations are classified in three spectral bands: (i) The secular variations: these fluctuations refer to the apparent linear increase in the length of the day. The longer period components of secular changes are produced by tidal friction, and internal sources such as changes in moment of inertia of solid Earth. It is also modified by the melting of ice [Chao, 1988] and the tidal dissipation torque. The tidal effect (the response of a system to fluctuations in net gravity) as generated due to the gravitational attraction of the Sun-Moon-Earth system is well recognized. The resultant tidal effect is quite dominant even for slight changes in the Earth's ellipticity. Consequently the Earth rotation changes and conserves the

Nonlinear Dynamics and Predictability of Geophysical Phenomena
Geophysical Monograph 83, IUUG Volume 18
Copyright 1994 by the International Union of Geodesy and Geophysics and the American Geophysical Union

angular momentum of the Earth. There are several tidal components in the LOD time series: the prominent ones are clustered around 9 to 27 days. [David, 1989] (ii) Irregular "decadal variations" (up to several milliseconds/century in the LOD): they are caused possibly due to angular momentum transfer between the Earth's solid mantle and fluid core. The transfer-effect causes slowing of the speed of the Earth's rotation resulting in lengthening of the day by about 0.0005 to 0.0035 seconds/century [McCarthy, 1991], and (iii) The unpredictable and most rapid "Non-tidal" variations on time scales ranging from days up to a few years: They have been shown to have amplitudes (up to about 1 millisecond in amplitude) and are largely of atmospheric origin. The comparative study of Earth orientation and Angular Atmospheric Momentum (AAM) has provided better understanding of the coupling between solid Earth, the ocean and the atmosphere [Dickey et al., 1986].

Until recently classical analyses of power spectra and correlation functions have been used to describe and classify the periodic/aperiodic characteristics of complex nonlinear behavior of Earth's angular rotation. No doubt, these statistical and spectral techniques are quite useful for analyzing processes having higher degree of freedom higher dimensions (>10). Such techniques, however, are not very suitable to provide meaningful information for a wide class of nonlinear coupled oscillations known as chaotic oscillations. Such chaotic oscillations exhibit broad bands and aperiodic spectral characteristics. It can be due to periodic deterministic motion arising due to nonlinearity and are extensively sensitive to initial conditions rather than stochastic behavior [Kaiser, 1990]. The study of LOD time series using nonlinear dynamical system theory is, therefore, appropriate for understanding of the coupled Earth's rotation-atmospheric oceanic processes.

THE LOD DATA AND FOURIER POWER SPECTRA

A fairly good number of data points is required to estimate the attractor dimension. The limited number of available geophysical/astronomical observations create difficulties in an accurate estimate of the attractor dimension using Grassberger and Procaccia (G-P) Algorithm, (1983a). However, the results of the original data are compared with the results obtained for the first order auto-regressive random model to test the stability of the results. A short span of accurate LOD time record has been analyzed using the nonlinear dynamical system theory of Grassberger and Procaccia (1983b). The original (UT_1–AT) data obtained from the Bureau International de l'Heure (BIH) (M. Feissel Personal Communication) is used here. The LOD data set used in the present analysis is the time derivative of UT_1–AT, and comprises 5328 points covering daily observations of roughly a 15-year-period (1976-1990) (Figure 1a). Figure 1a apparently depicts a quasi-periodic and stochastic time series of the original LOD data. Fourier power spectra of the data shown in Figure 1b exhibits noisy spectrum. It also shows broadband components and apparent spikes centered around annual, semiannual, 27, 13 and 9 days period. Such characteristics are well known for almost all the chaotic systems [Parker and Chua, 1989; pp 20].

CHAOTIC ATTRACTORS

All natural processes follow certain laws of physics in micro-to-macro time scales. The rotation/revolution of the Earth and planetary motion might broadly exhibit intrinsic order and harmony. However, a famous mathematician, H. Poincaré, of the late nineteenth century discovered that certain evolving mechanical systems governed by Hamilton's equations could display chaotic motions. Later, in 1963, the noted meteorologist, E.N. Lorenz found that even a simple set of deterministic first order coupled nonlinear differential equations can lead to chaotic trajectories. Most recently, Sussman and Wisdom (1988) have shown quantitative evidence that motion of the planet Pluto is chaotic. What do we mean here by "Chaos"? We are, in fact, not aware of any widely accepted definition of chaos. It might describe a notion of disorder/irregularity and unpredictability. According to Lorenz (1991), however, "chaos" means a physical phenomena that may appear to be fluctuating randomly but upon closer examination may possess considerable regularity. The LOD spectrum apparently exhibit certain intrinsic aperiodic components with some arbitrary components (Figure 1b). The observations indicate that dynamics of the LOD system may combine both stochastic and some regular components. It has been noted that the apparent regularity might be related to the inherent nonlinear dynamics in the interacting system but the arbitrariness probably stems from the sensitive dependence on initial conditions. The chaos, thus, adds a new quality of irregular stochasticity with some degree of determinism [Kaiser, 1990].

A certain classification of chaotic or non-chaotic system (or in other words random and/or deterministic components of a physical process) can be made using the embedding theory of nonlinear dynamical system [Takens, 1981]. The evolutionary fluctuations in the LOD exhibit the features of a dissipative dynamical system with possible attractors in phase space. If all the trajectories of evolutionary processes converge towards a "subset" of phase space, irrespective of their initial conditions, the subset or "submanifold" are called an "attractor" [Tsonis and Elsner, 1988]. An interesting discussion on chaos physics and attractors are given by Lauterborn and Parlitz (1988). Accordingly, the nature of attractor is given by its fractal dimension. For example, a physical process in two variables possessing a limit cycle has a periodic trajectory as its attractor and its dimension as unity. A quasi-periodic process exhibiting several non-harmonic frequencies may have a torus as its attractor. This may happen in a physical situation where the system oscillates with two or more incommensurable frequencies. In contrast, a "strange attractor" will have a non-integer dimension (Fractal dimension) and can be represented by more complex form in a phase space. It is characterized by chaotic trajectories and a broadening of the frequency spectrum. The fractal dimension of a chaotic system (strange attractor) indicates the number of ordinary differential equations for modeling the process, or in other words, it is indicative of the number of variables required for almost complete description of the dynamic system [Tsonis and Elsner, 1988].

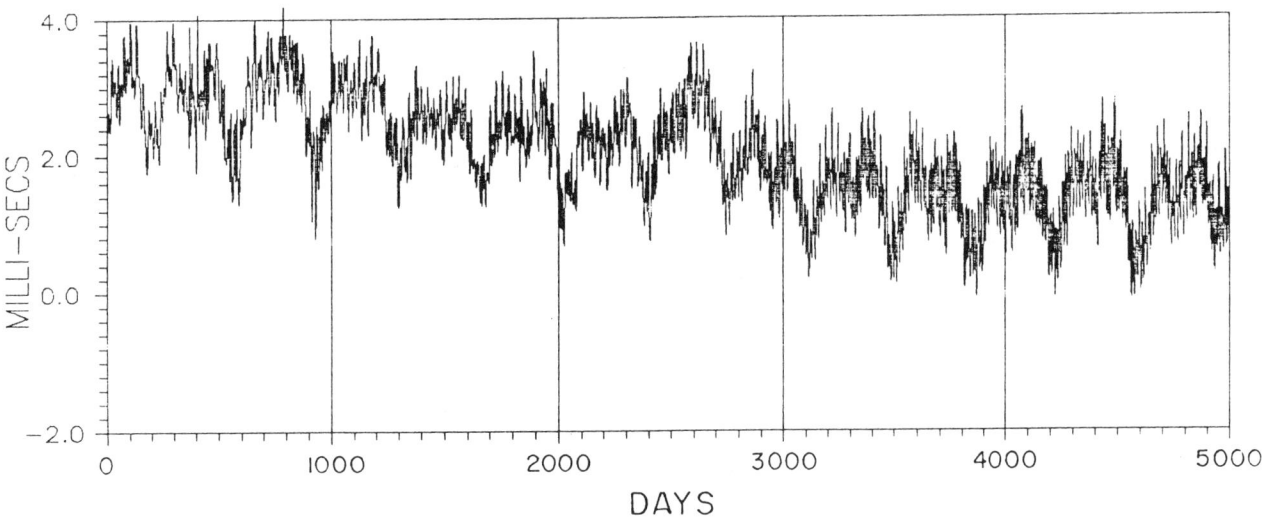

Fig. 1. (a) Time series showing fluctuations in the length of the day (LOD). (b) Fourier Spectrum of the record.

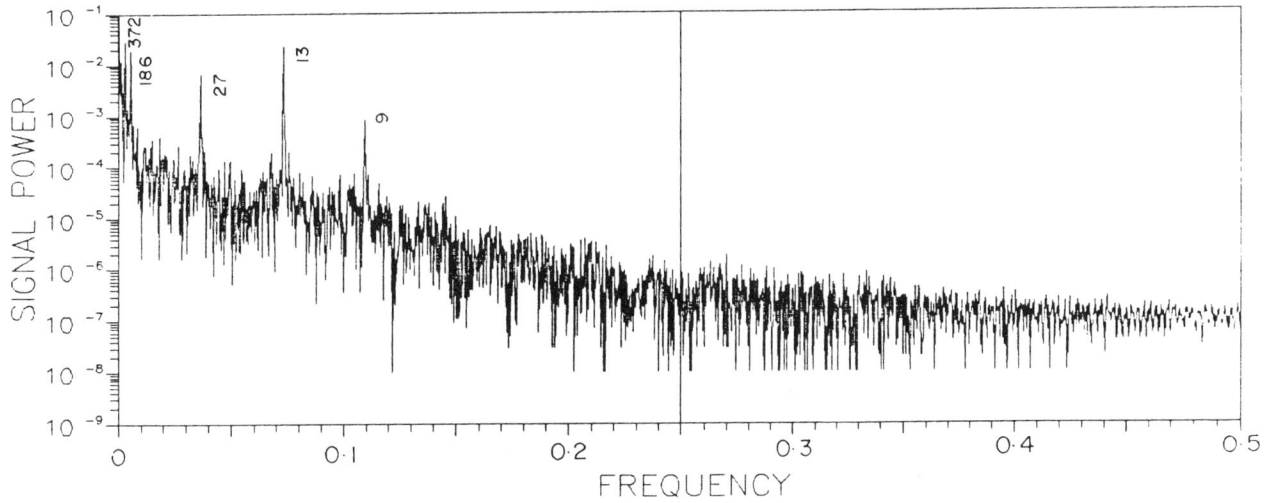

ALGORITHM FOR NONLINEAR TIME SERIES ANALYSIS

We often have access to time series of only a few variables. For a sufficiently coupled system, it is possible to use these variables to construct the phase space using the time delay embedding technique of Takens (1981). The characteristic quantities such as attractor dimension and entropy can then be found from the system's evolution in phase space.

Following Takens (1981) we construct an m-component "state" vector X_i from a time series $x(t)$ as:

$$X_i = \{x_1(t_i), x_2(t_i), ..., x_m(t_i)\} \quad (1)$$

where $x_k(t_i) = x[t_i + (k-1)\tau]$ and τ is an appropriate time delay (of the order of characteristic physical time scales). The distribution of state vectors in the reconstructed phase space is directly related to the dimension. A suitable distribution dependent quantity can be defined to examine its scaling with distance in phase space. The correlation integral [Grassberger and Procaccia, 1983a] for N vectors distributed in an m-dimensional space may be given as follows:

$$C_m(r) = \lim_{N \to \infty} \frac{1}{N^2} \sum_{i=1}^{N} \sum_{j=1, j \neq i}^{N} \theta(u) \quad (2)$$

where $\theta(U) = (r - |X_i - X_j|)$ and r is the distance between point pairs in phase space and the distances are taken in terms of Euclidean norm. The distance distribution function will obey a power-law scaling for small r values if the number of data points is large enough. This power-law relation for the attractor dimension 'd' is given by

$$d = \lim_{r \to 0} \frac{\log C_m(r)}{\log r} \qquad (3)$$

where d is the correlation dimension.

The correlation dimension is seen to converge (stabilize) to some definite value smaller than embedding dimension 'm' when the system is of deterministic nature with a low dimensional state space. The dimension of an attractor calculated from equation (3) is a lower bound on relevant number of state variables (degrees of freedom) needed to describe the steady-state behavior. The equality holding $d \geq m$ is indicative of a random or "noisy" system with many degrees of freedom.

IDENTIFICATION OF ATTRACTOR DIMENSION IN LOD TIME SERIES

The two dimensional projection of the trajectory is shown in Figure 2. Although it is evident from these figures that the trajectories are bounded, they do not clearly reveal a periodic pattern. Such behaviors are somewhat similar to chaotic pattern [Parker and Chua, 1989, pp 21]. Low dimensional physical system (with limited number of variables) like one presented in Figure (2) could possibly be modeled and predicted at least, in principle.

The power spectrum (Figure 1b) shows some sharp peaks centered around 27, 13 and 9 days in the subannual variation of the LOD. One might naturally suspect that apparent tones are related to solid Earth tides [Yoder et al., (1981)]. In order to demonstrate more convincingly the evidence for low dimensionality of chaotic pattern inherent in the nonlinear LOD time series, we have eliminated these tidal peaks using the harmonic elimination filter [Delache and Scherrer, 1983]. We then recompute the integral coefficients $C_m(r)$ according to equation (2), for a range of parameter values of r and embedding dimension m. As an example, the plot of values of $\log C_m(r)$ as a function of $\log(r)$ for filtered LOD data are shown in Figure 3a. It may be noted that as we proceed for calculation of $C_m(r)$, there is large fluctuation in $C_m(r)$. This occurs due to the fact that as the number of points decreased, the population pair at small scales gets more and more completely diffuse and masks the scaling region [Chouet and Shaw, 1991]. One should therefore be very careful in determining the dimension 'd' in these regions. We mention here that prior to construction of phase space and computation of integral coefficients, we look for the first minimum of autocorrelation function computed for the LOD series. It is required (although this is not unique) to construct the phase space and to calculate the integral correlation coefficients. Here the autocorrelation function of the LOD data drops to first minima at 6 (days) time units. We have constructed the phase space vector by using this value as the time lag.

Identification of the linear portion of this plot is crucial for calculating dimensions 'd' and to ensure that the power-law holds. The slope 'd' of the linear part of the curve for each correlation dimension is found by a least square method as shown in Figure 3b (with + marks) and compared with the slope obtained from unfiltered data sets (with × marks). Our result of the filtered time series

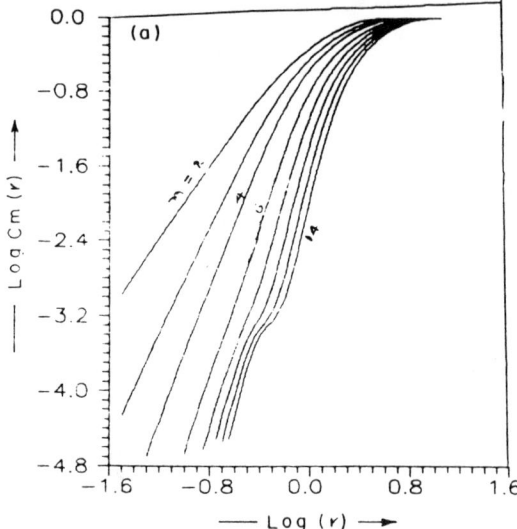

Fig. 3a. $\log C_m(r)$ versus $\log r$ for embedding dimensions (2–14) after filtering tidal components corresponding to 9, 13 and 27 days.

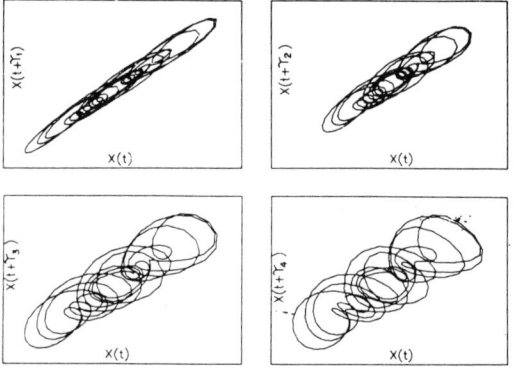

Fig. 2. Phase space projections of the LOD attractor. Data from Fig. 1a. In this example $\tau_1 = 1$ day; $\tau_2 = 2$ days; $\tau_3 = 3$ days and $\tau_4 = 4$ days.

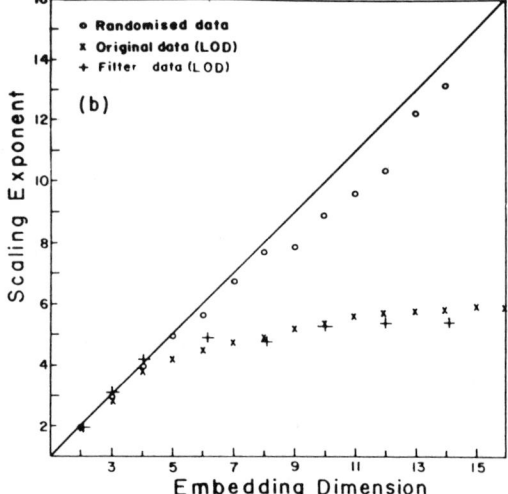

Fig. 3b. Comparison of scaling exponent versus embedding dimension (× original, + filtered and 0 random data).

shows low dimensionality and is consistent with our earlier results [Tiwari et al., 1992]. Figure 3b reveals that the computed slope increases with increasing values of m and then approaches a constant value for sufficiently high values of m. The slope vs. embedding dimension plot will be a straight line with constant slope of unity for a completely random process [Atmanspacher and Scheingraber, 1986]. Besides several other important factors (such as delay time, the choice of sampling and procedure of embedding a fixed number of points in higher and higher dimensions) controlling the attractor dimension, the amount the data points play critical role in precise estimation of 'd'. Because of the finite number of data available to us, it would be appropriate to compare these results with a random model to see how these actual results differ from a purely random one. Accordingly, we employ a nonlinear stochastic model [Fuller, 1976; pp 338], such as a first order auto-regressive random model of the form:

$$Y_t = \pi Y_{t-1} + \varepsilon_t \quad t = 1, 2, 3, ..., N \qquad (4)$$

where ε_t are normal independent random variables uniformly drawn from the interval (0,1). The maximum likelihood estimator is calculated from the LOD data. The results for the random model comprising the same number of data points as the original data set are also superimposed in Figure 3b by open circles. It is evident from Figure 3b that estimated dimension from the first order auto-regressive random signal closely matches the white noise process. However, the correlation dimension computed for the LOD data clearly indicates the existence of a deterministic component. The slope of $\log C_m(r)$ vs. $\log r$ curves is independent of embedding dimension and becomes saturated for higher embedding dimensions. The saturating value of the information dimension indicates the existence of a low dimensional attractor. Figures 3a and 3b apparently depict the value of 'd' in the range of 5 to 7.

THE KOLMOGOROV ENTROPY (K_2)

The Kolmogorov entropy (K_2), which measures the rate of loss of information, or difference of evolution between almost identical initial conditions, is a lower bound of the entropy. When K_2 entropy is finite and non-zero the system is chaotic whereas if K_2 is infinite the system is random (nondeterministic). The inverse of this entropy is a time scale over which we can accurately predict the behavior of the system [Vassiliadis, et al. 1990]. One can easily calculate the K_2 entropy from equation (2) [Grassberger and Procaccia, 1983b] from the correlation integrals

$$K_2 = \lim_{r \to 0} \frac{1}{\tau} \ln \left[\frac{C_m(r)}{C_{m+1}(r)} \right] \qquad (5)$$

where τ is the sampling rate.

Using equation (5), one can calculate the Kolmogorov entropy (K_2) for different values of embedding dimensions. Figure 4 shows the K_2 entropy for embedding dimensions (1–15). The convergence value is about 0.13/day. It is an estimate of the Kolmogorov entropy of the LOD signal and is indicative of chaotic behavior of the LOD. Although our results indicate the possible existence of low dimensional chaotic attractor, it must be reconfirmed in future with the

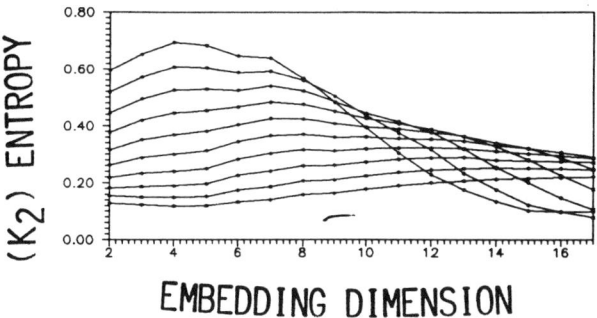

Fig. 4. The entropy (K_2) versus embedding dimension (1–15).

availability of a large LOD data set. It may not be prudent at this stage to confirm the chaotic nature of the LOD dynamics by the limited number of data points.

DISCUSSION

Several earlier studies of various climatic and meteorological records [Fraedrich, 1987; Essex et al., 1987; Henderson and Wells, 1988] have also shown possible evidence for a chaotic attractor with an apparent dimension in the range of 5 to 7. In a recent investigation Malinetskii et al. (1990) have used six monthly averaged data as day long term variations for 1656–1984 to obtain a lower bound of the dimensions of the attractor of a dynamic system with the Earth's angular velocity as a variable. Their analysis has shown evidence for a low dimensional attractor of about 1.3 ± 0.3 suggesting that the LOD data for the above period can be explained by using models with a few degrees of freedom. On the other hand, the LOD data analyzed here are of short span with daily variations. The present analysis reveals dimensionality in the range of 5 to 7. It might be interesting here to indicate interacting physical processes which might contribute to such a low dimensionality.

Ambiguous spectral characteristics are common in chaotic systems [Parker and Chua, 1989; pp 20]. Some such apparent regularities in the physical system sometimes exhibit one of the features of chaotic dynamics known as "periodic doubling chaos" [Lauterborn and Partitz, 1988]. Some apparent periodicities, visually evident in Figure 1b, might be indicative of "periodic doubling chaos" [Tiwari, 1991]. At present such arguments are, however, merely speculation and are being tested against "Logistic map." The reason for a low dimensional strange attractor may possibly be associated with short period atmospheric momentum changes in the higher frequency range due to thermally forced seasonal and coupled air-sea level oscillations [Rosen and Salstein, 1983; Eubanks et al., 1985; and Morgan et al., 1985]. Thus, a low dimensional attractor in the LOD exhibits a near equilibrium state [Nicolis and Nicolis, 1984] and coupling with self-regulating atmospheric and ocean processes.

We ponder over some of the following well-known possibilities discussed in detail by Dickey et al. (1986). (i) Short-term LOD variability are mainly dominated by the changes in the atmospheric angular momentum (AAM). The momentum contains a large sea air cycle dominated by annual and semiannual harmonics. The seasonal cycles of global weather patterns (e.g. Monsoon) also cause seasonal LOD changes and exhibit low dimensional strange attractor in a range of 6–7 [Tsonis and Elsner, 1988]. The AAM annual cycle is mostly due to changes in the mid-latitude westerlies (i.e., propagating eastward) including the subtropical jet streams (current of strongest wind in each hemisphere encircling the globe) mostly coming from high velocity winds at or near the 25 millibar level [Rosen and Salstein, 1983]. (ii) Superimposed on the seasonal cycle is the irregular 50 day oscillation which varies in period for 40–60 days. Most of the nonseasonal changes occur in regions from 10 to 25 degrees south and 20 to 35 degrees north [Rosen and Salstein, 1983]. (iii) There may be a weak relationship between El Niño (a quasi periodic cycle of weak trade winds and warm, wet conditions in the eastern tropical Pacific) events and the rapid (40 to 60) days fluctuation in the LOD. (iv) The zonal component of the angular momentum (using wind and pressure data) suggests that the angular momentum transfer between the Earth and its atmosphere probably accounts for most of the observed variations [Hide et al., 1980]. Most of the annual monsoon-like imbalance can be attributed to the atmospheric wind and oceanic circulation to contribute possible changes in the LOD.

Summary

The state space of the LOD variation (axial rotation of the Earth) were reconstructed by using the nonlinear technique described briefly above from the time series data. The system in the reconstructed state space is seen to evolve on a fractal set within a range of 5–7. The results indicate two aspects: (i) The dynamics of the LOD system is chaotic and (ii) it has a low dimension attractor. Thus, the annual air circulation effects can be modeled by at least 5 to 7 independent variables. However, both the possibilities should be reconfirmed with more LOD data points. The information dimension of the phase space reported here indicates near equilibrium processes in the redistribution of its angular momentum in the coupled ocean and atmospheric processes. These results provide interesting inferences to derive certain invariant quantities needed to model the deterministic part of the nonlinear dynamics of the LOD. It would be interesting to identify the appropriate physical variables and to search possible forms of equations for the evolution of the variations in the length of the day. Such efforts would require proper understanding of various physical processes and nonlinear analyses. Low dimensionality of the LOD attractor is quite encouraging to develop an approach to formulate a mechanism for short term LOD variability.

Acknowledgements. The data used in this work were kindly provided by Dr. M. Feissel. We are grateful to her. We are also grateful to an anonymous reviewer for his constructive comments and criticisms to improve the manuscript. Editors of this volume Professors Andrei M. Gabielov and W.I. Newman are gratefully acknowledged for their keen interest and timely information to include this contribution. Thanks are also due to Mr.Ch. Ramaswamy for help in the preparation of the manuscript. The Director, NGRI, is gratefully acknowledged for his kind permission to publish the manuscript.

References

Atmanspacher, H., and H. Scheingraber, Deterministic chaos and dynamical instabilities in a multimode cw dye laser, *Phys. Rev.*, *A 34*, 253–263, 1986.

Chao, B.F., Interannual length of day variation with relation to the southern oscillation El Niño, *Geophys. Res. Letts.*, *11*, 541–544, 1984.

Chao, B.F., Correlation of interannual length of day variation with El Niño/southern oscillation, 1972–1986, *J. Geophys. Res.*, *93*, 7709–7715, 1988.

Chouet, B., and H.R. Shaw, Fractal properties of Tremor and gas piston events observed at Kilauea Volcano, Hawaii,*J. Geophys. Res.*, *96*, no. b6, 10, 177–10,189, 1991.

David, E. James, (Ed.), *The Encyclopedia of Solid Earth Geophysics*, van Nostrand Reinhold company, New York, 237, 1989.

Delache, P., and P.H. Scherrer, Detection of solar gravity mode oscillations, *Nature*, *306*, 651–654, 1983.

Dickey, J.O., High accuracy Earth rotation and atmospheric angular momentum, in *Earth Rotation and Unsolved Problems*, edited by A. Cazenave, pp. 137–162, Reidel Pub. Comp., Nato ASI Series C. (Mathematical and Physical Analysis, vol. 187), 1986.

Dickey, J.O., T.M. Eubanks, and R.Hide, Interannual and decade fluctuations in the Earth rotation: Paper presented at 19th General Assembly of the IUGG, Vancouver, B.C. Canada, August 18–19, 1987.

Essex, L., J. Lookman, and M.A. Nerenberg, The climate attractor over short time scales, *Nature*, *326*, 64–66, 1987.

Eubanks, J.M., J.A. Steppe, J.O. Dickey, and P.S. Callahan, A spectral analysis of the Earth's angular momentum budget: *J. Geophys. Res.*, *90*, 5385–5404, 1985.

Fraedrich, K., Estimating the dimensions of weather and climate attractors, *J. Atmos. Sci.*, *44*, 722–728, 1987.

Fuller, W.A., *Introduction to Statistical Time Series*, John Wiley and Sons, New York, 1976.

Grassberger, P., and I. Procaccia, Characterization of strange attractor,*Phys. Rev. Letts.*, *50*, 346–349, 1983a.

Grassberger, P., and I. Procaccia, Estimation of the Kolmogorov Entropy from a chaotic signal, *Phy. Rev.*, *A 28*, 2591, 1983b.

Henderson, H. W., and R. Wells, Obtaining attractor dimensions from meteorological time series, *Advances in Geophysics*, *30*, 205–237, 1988.

Hide, R., T. Burch, L V. Morrison, D.J. Shea, and A.A White, Atmospheric angular momentum fluctuations and change in the length of the day, *Nature*, *206*, 114–117, 1980.

Kaiser, F., Nonlinear dynamics and deterministic chaos. Their relevance for biological functions and behaviour, in *Geocosmic Relations, The Earth and Its Macro-Environment*, edited by G.J.M. Tomassen et al., pp. 315–320, The Netherlands, Pudoc., 1990.

Lauterborn, W., and U. Parlitz, Method of chaos physics and their application to acoustics, *Jr. of Acoustic Soc., Am.*, *84*, 1975–1993, 1988.

Lorenz, E.N., Deterministic non-periodic flow, *J. Atmos. Sci.*, *20*, 132–141, 1963.

Lorenz, E.N., The general circulation of atmosphere a evolving problem. *Tellus*, 8–16, 1991.

Malinetskii, G.G., A.B. Patapov, S.M. Gizzatulina, A.A. Ruzmaikin, and V.D. Rukavishnikov, Dimension of geomagnetic attractor from data on length of day variations, *Phys. Earth, Planet. Inter.*, *59*, 170–181, 1990.

McCarthy, D.D., Astronomical time, *Proceeding of the IEEE*, *79*, 7, 915–920, 1991.

Morgan, P.J., R.W. King, and I.I. Shapiro, Length of the day and atmospheric angular momentum, A comparison for 1981–1983, *J. Geophys. Res.*, *90*, 12645–12652, 1985.

Nicolis, C., and G. Nicolis, Is there a climatic attractor, *Nature*, *311*, 529–532, 1984.

Parker, T.S., and L.O. Chua, *Practical Numerical Algorithms for Chaotic Systems*, Springer-Verlag, New York, 1989.

Rosen, R.D., and D.A. Salstein, Variation in atmospheric angular momentum on global and regional scales and the length of the day, *J. Geophys. Res.*, *88*, 5451–5470, 1983.

Sussman, G.J., and J. Wisdom, Numerical evidence that the motion of Pluto is chaotic, *Science*, *241*, 17–23, 1988.

Takens, F., *Detecting strange attractors in turbulence in dynamical systems and Turbulence, Warwick 1980, Lecture Notes in Mathematics*, edited by D. Rand and L.-S. Young, pp. 366–381, Springer-Verlag, New York, 1981.

Tiwari, R.K., Chaotic attractor in nonlinear fluctuations of length of day (LOD) variations. Invited paper presented at the Workshop 'On Fractal and Chaos' held in Department of Mathematics and Computer Sciences at the University of Hyderabad, Feb. 21–23, 1991. Tiwari, R.K., J.G. Negi, and K.N.N. Rao, Attractor dimension in non-linear fluctuation of length of day time series, *Geophys. Res. Letts.*, *19*, 909–912, 1992.

Tsonis, A.A., and J.B. Elsner, The weather attractor over very short time scales, *Nature*, *333*, 545–547, 1988.

Vassiliadis, D.V., A.S. Sharma, T.E. Eastman, and K. Papadopoulos, Low dimensional chaos in magnetospheric activity from AE time series, *Geophys. Res. Letts.*, *17*, 1841–1844, 1990.

Yoder, C.F., J.G. Williams, and M.E. Park, Variations of Earth rotation, *Jr. Geophys. Res.*, *86*, NB2, 881–891, 1981.

R.K. Tiwari, J.G. Negi, and K.N.N. Rao, National Geophysical Research Institute, Hyderabad 500 007, India.

Self-Organized Criticality: Consequences for Statistics and Predictability of Earthquakes

PER BAK, KIM CHRISTENSEN, AND ZEEV OLAMI

Brookhaven National Laboratory, Upton, New York

The concept of self-organized criticality provides a natural, robust explanation of the statistics of earthquakes, including the Gutenberg-Richter law for the distribution of earthquake magnitudes. The dynamics is "at the edge of chaos" with algebraic, not exponential, divergence of small uncertainties. Temporal clustering of big earthquakes arises because of an underlying fractal structure of correlated regions. Scaling laws suggest that the statistics of large events can be inferred from the statistics of the much more numerous small events.

INTRODUCTION

The frequency of earthquakes versus magnitude exhibits a logarithmic dependence over many decades, known as the Gutenberg-Richter law [Gutenberg and Richter, 1956]. Figure 1 shows data for earthquakes in the New Madrid zone collected by Johnston and Nava [1985]. The magnitude is proportional to the logarithm of the energy release, or the seismic moment, so the G-R law indicates that these quantities obey a power law distribution, with exponent $1+\beta$, and $\beta \approx 0.8$. Measurements of the exponent β vary from place to place, so the exponent appears to be non-universal. The spatial distribution of earthquake occurrence seems to be fractal. The distribution of aftershocks follows Omori's law, another power law. Of course, such power laws can not extend to infinite energies; there has to be a cutoff somewhere: if for no other reason, then because of the finite size of the earth. From measurements over a finite period, say the last 60 years, all we can say is that this cut-off exceeds earthquakes of size 9.

These facts seem to imply that we are dealing with a kind of critical phenomenon, because power laws for temporal and spatial correlation functions are the hallmark of systems at a critical point for a continuous phase transition. Indeed, the mathematician Vere-Jones [1977] demonstrated that in principle a power law (with an exponent $\beta=0.5$) could be formally explained by thinking of an earthquake as a critical chain reaction, starting from a single rupture event. At each branching point the probability of doubling the activity must be precisely balanced by the probability of death of the activity (Figure 2). But why should the chain reaction be exactly critical, since with unit probability chain reactions are either sub-critical, in which case large events would be exponentially unlikely, or super-critical, in which case the activity would explode exponentially?

A few years ago one of the authors, in collaboration with Chao Tang and Kurt Wiesenfeld [Bak, Tang, and Wiesenfeld, 1987, 1988a,b; Wiesenfeld, Bak, and Tang, 1989; for a review see Bak and Chen, 1991] demonstrated that slowly driven dynamical systems, with many degrees of freedom, naturally self-organize into a critical state, with avalanches of all sizes obeying power law statistics. The critical state is an attractor for the dynamics. The phenomenon is deterministic and robust with respect to noise and inhomogeneities. Large catastrophic events are intrinsic, unavoidable properties of those large interactive systems, and no external or internal cataclysmic force is necessary. Thus, in contrast to equilibrium physics where criticality is the exceptional case, in non-equilibrium physics criticality could be the typical state of matter.

This opens up for an entirely new view on many sciences, including Biology [Raup, 1986], where catastrophic events have occurred intermittently in the past, most notable the extinction of the dinosaurs 50 million years ago, Economics, with power law tails for the fluctuations on Wall Street and the distribution of price variations [Mandelbrot, 1963], and Geophysics, not only in relation to earthquakes but also for volcanic eruptions [Diodati, Marchesoni, and Piazzo, 1991].

The obvious applicability of self-organized criticality to earthquakes was immediately and independently pointed out by several authors [Bak and Tang, 1989; Ito and Matsuzaki, 1989; Sornette and Sornette, 1989; Carlson and Langer, 1989]. The initial models were quite crude local sand-pile type models, but nevertheless served to illustrate the viability and the robustness of the idea. Subsequent rupture models [Chen, Bak, and Obukhov, 1991; Xu, Bergersen, and Chen, 1991] included the long-range

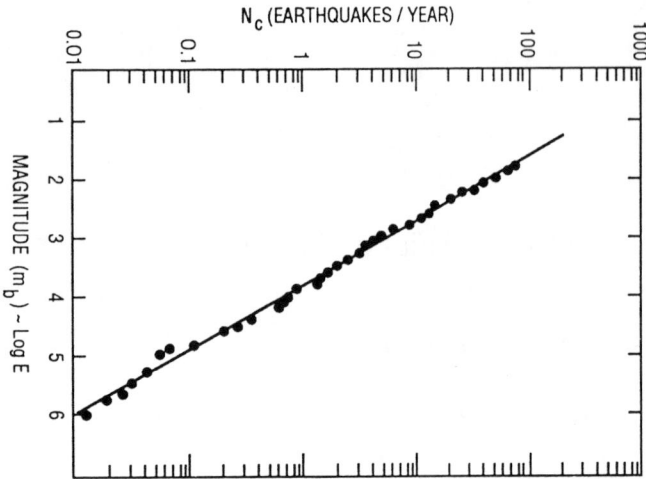

Fig. 1. Cumulative distribution of earthquakes in the New Madrid zone during the period 1974-1983. The data were collected by Johnston and Nava [1985].

redistribution of elastic forces following rupture. A thorough analysis of two-dimensional models derived from Burridge-Knopoff models [1967] of blocks connected with springs has been carried out by two of us, in collaboration with Hans Jacob Feder [Olami, Feder, and Christensen, 1992; Christensen and Olami, 1992a,b]. In contrast to the original models, these latter models do not invoke any conservation laws for force redistribution, thus removing one artificial artifact. The striking picture which emerges from the SOC theories is that the crust of the earth on which we are living operates at a perpetually critical state, always at the verge of collapse.

In the following we shall discuss results based on the study of SOC models of earthquakes, and point out the consequences for our ability of predicting earthquakes, in particular their statistical properties. Results for model calculations are discussed in the context of observations. Bear in mind that a more or less complete understanding of a physical phenomenon does not necessarily allow us to predict the future, as is the case in quantum mechanics, and for chaotic systems. Our motivation is a desire to understand, rather than to predict or prevent specific events. We find, however, that earthquakes is not a chaotic phenomenon, so at least there are no fundamental dynamical principles preventing us from predicting earthquakes.

We suggest that scaling laws be exploited to deduce the statistics for the few large earthquakes from the statistics of the much more numerous small earthquakes. Specifically, it will be argued that large earthquakes are clustered rather than periodic, contrary to popular belief (but in agreement with observations), and that the occurrence of large characteristic events are illusions based on some peculiar features of power law distributions. The danger of such mirages in fractal phenomena has been pointed out by Mandelbrot [1963, 1982].

SOC SPRING-BLOCK MODELS OF EARTHQUAKES

The idea of self-organized criticality, as applied to earthquakes, may be visualized as follows: Think of the crust of the earth as a collection of tectonic plates, being squeezed very, very slowly into each other. In the beginning of our geological history, maybe the stresses were small, and there would be no large ruptures or earthquakes. During millions of years, however, the system evolved into a stationary state where the build-up of stress is balanced in average by the release of stress during earthquakes. Because of the long evolutionary process, the crust has "learned", by suitably arranging the building blocks at hand into a very balanced network of faults, valleys, mountains, oceans and other geological structures, to respond critically to any initial rupture. The result of this self-organization process is in sharp contrast to any network of faults that one might set up by construction or engineering, which would certainly not be critical. We do not know how it all started, but that is not important for our arguments: the self-organized critical state is an attractor of the dynamics which will be reached eventually irrespectively of the initial conditions.

It makes no sense to separate the dynamics from the statics. It is not productive to think of earthquakes as being generated by "pre-existing faults". One can trivially explain the G-R law by assuming a fractal distribution of faults with a power-law distribution of characteristic fault sizes, but that leaves us with the equally difficult problem of explaining the dynamical origin of that distribution. Popularly phrased, one must take a holistic view of the situation. What appears to be a static configuration of large faults in a human lifetime merely constitute a snapshot of a slow ongoing geological process that has been hundreds of millions of years underway. During that period, faults have come and gone. The dynamics of the fault structure and the Gutenberg Richter law must be produced within a unified picture. The SOC models simulate the long term dynamics of the crust. In order to represent a realistic view of geophysics, the models must be robust, or adaptive, in the sense that if the physical properties were changed, or if noise were added, the system would reorganize during a transient period and become critical again. This is indeed the case for SOC models of earthquakes.

We want to study the simplest possible models which contain the essential physics of earthquakes. While there has been studies of three-dimensional crack-propagation models with slightly more realistic long-range redistribution of elastic forces following rupture, simple local models are probably more instructive, and certainly much more amenable to numerical and analytical study. We stress that we don't think of the Gutenberg-Richter law as originating from a single fault, which must necessarily have a characteristic energy depending on the size of that fault: our models

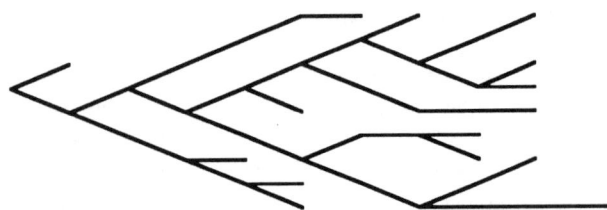

Fig. 2. Rooted tree generated by a critical branching process. Each branch indicates a rupture event, and the complete tree represents a single earthquake.

are "toy" models supposed to illustrate the principle of global organization of the crust of the earth.

Consider a two-dimensional lattice of interacting blocks. The initial block structure merely represents a discretization of the space in much the same way as the lattice in lattice gauge theories of particle physics. The block size does not represent an intrinsic length scale in the problem. On each block, at sites (i,j), acts a force $F_{i,j}$ in the general direction of motion. In the beginning $F_{i,j}$ may assume some random, small value. The initial state is not important for the long term dynamics. Let the force increase uniformly by a very small amount per unit time; this simulates the slow driving by the tectonic plate motion, or whatever force is driving the system. Eventually, the force at some site (i,j) must exceed a critical threshold value F_C for rupture. The critical force may either assume the same value at all sites, or be randomly distributed. The initial rupture is simulated by updating the forces at the critical site and the sites of the neighbors at (i,j±1) and (i±1,j). There are several possibilities for defining those rules:

a)
$$F_{i,j} \rightarrow F_{i,j} - F_C$$
$$F_{nn} \rightarrow F_{nn} + \alpha F_C$$

or b)
$$F_{i,j} \rightarrow 0$$
$$F_{nn} \rightarrow F_{nn} + \alpha F_C$$

or c)
$$F_{i,j} \rightarrow 0$$
$$F_{nn} \rightarrow F_{nn} + \alpha F_{i,j}$$

These equations represent the transfer of force to the neighbors. The updating of all sites is done in parallel. The transfer may cause the neighbors to be unstable and a chain reaction to take place. This chain reaction is the earthquake. The equations are completely deterministic. We are not dealing with a noise-driven phenomenon; on the contrary the physics turns out to be stable with respect to a small noise, i. e. noise is irrelevant. After a finite number of updatings, the forces on all sites will become sub-critical, and the earthquake stops. The system is then quiet until the force at some other location exceeds the critical value and a new event is initiated. The process is repeated again and again. One observes that the earthquakes become bigger and bigger for a long transient period. Eventually the growth stops: the crust has self-organized into a stationary state. At this point one may start measuring the seismic moment of earthquakes as defined by the total number of rupture events following a single initial rupture. A histogram similar to that in figure 1 for real earthquakes can be constructed.

The difference between the definitions a), b) and c) may seem subtle and irrelevant since one might expect $F_{i,j}$ to be not far from F_C at rupture, and certainly, by definition, is identical to F_C at the initial triggering instability. Actually, this difference turns out to be essential. In the continuous version of the BTW model for self-organized criticality (Bak, Tang, and Wiesenfeld, 1988b; Bak and Chen, 1991a) the equations a) were applied, with $\alpha=1/4$, so the force was conserved. However, as soon as α deviates from 1/4, the criticality gradually disappears, with a decreasing cut-off for large earthquakes. It turns out that there is no reason that the force be conserved for real earthquakes, so the model is not robust enough.

Recently, it was accidentally discovered by Feder and Feder [1991] that the situation b) with $\alpha = 1/4$ allowed for some criticality. Note that this model is non-conservative because sometimes the value at the critical site is larger than the threshold value F_C so the reduction of force at that site exceeds the amount transferred to the neighbors. Olami, Christensen and Feder realized that model c) could be directly related to earlier spring models studied for example by Burridge and Knopoff [1967], with the value of α directly related to the elastic parameters. The criticality in this case prevails for values of α down to 0.05, with only 20% conservation. This came as a surprise since there was at that time a widespread belief that the lack of conservation would spontaneously generate a length scale, i. e. a "characteristic earthquake size". It turns out that criticality occurs generically: it is almost independent of the details of the toppling rule. In fact, the situation a) is the one which is special in the sense that its rules do not induce correlation between toppling sites.

Figure 3 shows the distribution of earthquakes for $\alpha = 0.20$. The straight line on the log-log plot indicates a power law: the system has self-organized into the critical state. The slope of the line corresponds to a β value of 0.8. The slope turns out to depend on the degree of dissipation, $(1-4\alpha)$, so there is no universality of the exponent β in the non-conservative case. One should not look for unique b-values in nature. Indeed different b-values are observed in different regions of the world.

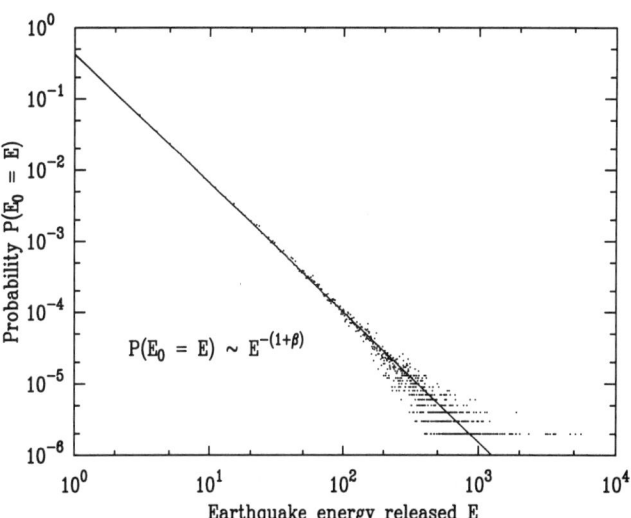

Fig. 3. The distribution function for the energy of earthquakes in our model. This graph represent a total of half a million earthquakes on a square lattice of linear size 100, with $\alpha=0.20$.

Note the single scattered events for values of E greater than the one where n(E)=1, i. e. the point where the straight line crosses the line n=1 (2×10^{-6} in the plot). This is due to the fact that the power law distribution function has the peculiar property that the average size of earthquakes diverges, so that a sampling from the distribution function never converges to the distribution function itself. The difference amounts to almost one full order of magnitude, and can not, even in principle, be eliminated by better statistics which would simply shift the problem to larger earthquakes. We are condemned to accept the fact that the statistics is poor precisely for the events that we are most interested in, namely the large events responsible for the highest energy release, and the most damage.

The observation of such single, seemingly atypically large earthquakes, has led to the concept of "characteristic earthquakes" not given by the G-R distribution (see Scholz, 1991) The size of these characteristic earthquakes is merely a consequence of the finite duration of the observation, typically something like a human lifetime. The slow logarithmic dependence of the "typical largest earthquake" on the observation period might lead to the belief that the time-scale, and the magnitude of those events are significant. Common sense, however, indicates that a human life time can play no role in a geophysical phenomenon such as earthquakes. Were we to live a million years, we would probably observe "characteristic" earthquakes of magnitude 13. Kagan [1992] has pointed out from analyzing actual earthquake catalogs that characteristic earthquakes indeed seem to be statistical artifacts. We shall see that those observations may actually be related to earthquake clustering.

Figure 4 illustrates the slow nature of the self-organization process. The running average of earthquake sizes vs. time is shown

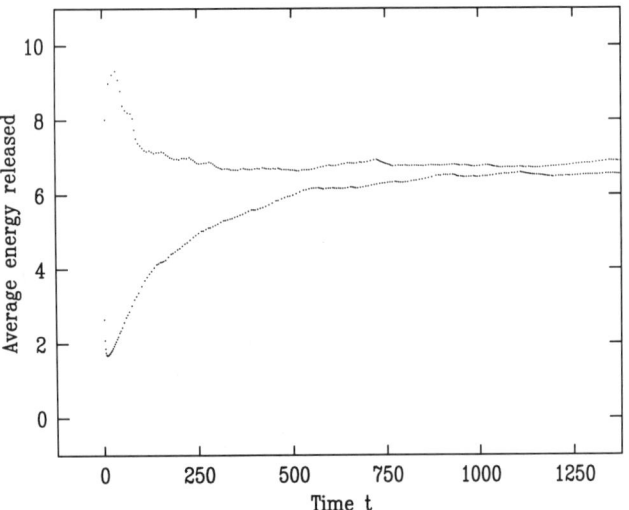

Fig. 4. Average size of earthquakes as a function of time during the self-organization process for a system of size L=70, conservation level $\alpha=0.20$, The lower graph represents the slow growth of the average for an initial random lattice. The growth rate is a measure of the correlation length in the system. Notice that the initial rise is linear. The saturation is an effect of the finite system size. The upper graph is data for an initially correlated lattice.

Fig. 5. A sequence of earthquakes for a 35x35 system with $\alpha=0.20$. The upper sequence shown is the occurrence of earthquakes with energy greater than or equal to 20. The lower sequence is for the occurrence of earthquakes with energy larger than 450. The time intervals are scaled so that the densities of events are the same. Note the clusters of "characteristic earthquakes". The life-time of the clusters may represent the active lifetime of an individual fault or fault structure.

in the lower curve, starting from a random uncorrelated configuration of forces. The growth will continue indefinitely, limited only by the size of the system. During the self organization process a time dependent cutoff will be seen in the size of the earthquake distribution. The upper curve shows the running average starting from a state which has already had time to reach the steady state. The initial fluctuations in the curves can be shown to be due to the fact that the exponent of the power law distribution is less then 2.

When comparing with real earthquake statistics, we assume that the crust of the earth has had sufficient time to complete the self-organization process. The power law distribution of earthquakes stems from the fractal nature of the SOC state, with correlated regions ranging over all length scales; those correlated regions, generated by the long term dynamics, are the equivalent of the active faults, or fault segments, in real earthquakes. The fault structure changes on large geological time-scales. The long range SOC models (Chen et al [1989] and Xu et al [1991]) produce a geometry which looks much more like a real fractal-like arrangement of two-dimensional faults in a three dimensional matrix.

Temporal Correlations

A very intriguing question of earthquakes is the temporal correlations between earthquakes. In figure 5 we show two temporal sequences of earthquakes derived from model c) with $\alpha=0.20$ for earthquakes with energy larger than 20 and 450, respectively, for a 35x35 system. It is evident that the two time sequences are dramatically different. The sequence for small earthquakes seems to be random, while for the large earthquakes the distribution is highly clustered. Also, the centers of the clustered earthquakes are generally correlated in space.

A possible measure for the observed temporal clustering is the coefficient of variation, $C_V(E)$. It is defined as the ratio between the square root of the variance of the temporal intervals, $Var(t)_E$, and the average interval between earthquakes with energy larger than E, $<t>_E$. For a random signal the distribution function is simply an exponential function yielding $C_V(E) = 1$. For a periodic signal $C_V(E) = 0$ while clustered earthquakes will produce $C_V(E) >$

1. We have measured the coefficient of variation for the conservative model with α=0.25 and for a non-conservative model with α=0.20. No correlations are seen between earthquakes in the conservative model, $C_V(E) \approx 1$. In the non-conservative model we see a clustering effect for large earthquakes, see figure 6, while random behavior is observed for small earthquakes in accordance with figure 5. The decrease in the coefficient of variation for very large earthquakes is a finite-size effect, related to the cutoff in the energy distribution. The same kind of temporal correlations is seen in real earthquakes. Small earthquakes seem to be uncorrelated [see figure 6 in Johnston and Nava, 1985]. Large earthquakes display strong clustering [Kagan and Jackson, 1991]. If one were to perform the same types of calculation for small subsystems, correlations between smaller events should eventually appear. This implies that it might be very useful to study smaller events in order to get more understanding of the larger events: one might predict the statistics of large earthquakes in large regions by scaling properly the statistics of smaller earthquakes in small regions.

The self-organization process of earthquakes creates big correlated strain structures which are responsible for the occurrence of large earthquakes. Usually those correlated structures do not disappear after a shock has occurred. Because of strain dissipation during the shock the strain in a correlated areas drops to a lower value. The same correlated areas have a relatively large probability to be activated again after a short loading time which is defined by the degree of non-conservation. The system remembers its past. This is the basic explanation for the spatio-temperal clustering of large earthquakes. It should be noted that the average time interval between large earthquakes is very large (proportional to E^β) while the characteristic time between clustered earthquakes is simply the loading time of the strain. Those two time scales are very different.

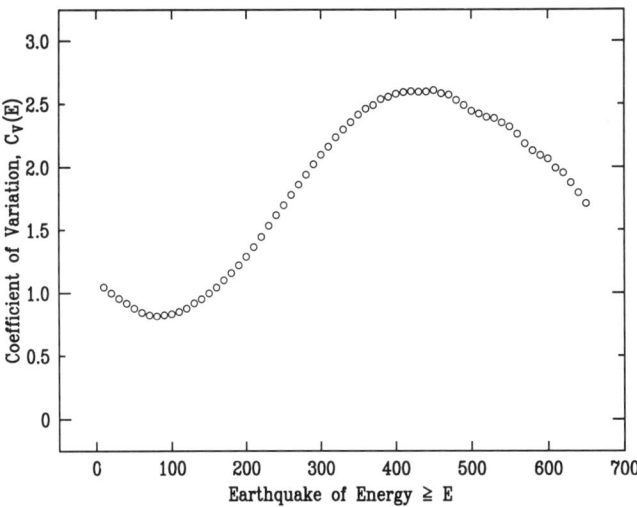

Fig. 6. The coefficient of variation $C_V(E)$ as a function of the energy released during an earthquake. The results are for α=0.20, L=35 and open boundary conditions. For the smaller earthquakes $C_V(E)<1$. The larger earthquakes are characterized by clustering, $C_V(E) >1$.

Thus, *locally* one observes a clustering of earthquakes of a certain magnitude related to a "pre-existing" correlated region, which might take the shape of a "fault". The system's memory is encoded into the fault structure. However, the cluster survives only for a certain time, i. e. the time scale for which the fault remains active; that is the interval between clusters in figure 5; this is long compared with the interval between large earthquakes, but short on a geological time scale. If one averages over long time, or over a large enough geographical area, the correlations disappear and the Gutenberg-Richter law is recovered. This was actually observed by Scholz in his study of earthquakes in Alaska (Scholz, 1991): data from individual "rupture zones" appear to include characteristic earthquakes, but when averaged over the whole zone the earthquakes obeyed the Gutenberg-Richter law.

ARE EARTHQUAKES CHAOTIC?

There has been a good deal of speculations that many complex phenomena in Nature are chaotic. In chaotic systems, a small uncertainty δ of the initial state of the system grows exponentially with time, $\delta = e^{+\lambda t}$ where λ is known as the Lyapunov exponent. This exponential growth makes the behavior of the system unpredictable at times larger than a characteristic time 1/λ. But critical systems have no characteristic time scale so how can they be chaotic?

Within the simple uncorrelated chain reaction picture the question has a simple answer. Compare the system with another critical system which initially is slightly different, for instance by starting with slightly different values of the forces. This causes a number of "mistakes" in the branching process illustrated in figure 2. Some sites which branch in the original model die, and vice versa. The number of mistakes grows linearly in time. The difference in the state of the two systems is simply the accumulated number of mistakes, which grows quadratically with time, $\delta = at^2$, not exponentially. In the more general case, where the branchings are correlated, one expects the divergence to be given by a different power law. Indeed, this is what is found in numerical studies of earthquake models [Bak and Chen, 1992; Chen, Bak, and Obukhov, 1991] and other self organized critical phenomena. Our conclusion based on the criticality indicated by the Gutenberg-Richter law and numerical simulations must be that earthquakes are not chaotic, so the structure of the dynamical equations does not in itself prevent earthquake prediction. Actually, the concept of self-organized criticality complements the concept of chaos wherein simple systems with a small number of degrees of freedom can display quite complex behavior.

The idea of self-organized criticality applies not only to earthquakes, but probably to most phenomena in Geophysics where power laws such as those characterizing spatial fractality have been observed. In particular, the intermittent nature of volcanic eruptions share many of the statistical features of earthquakes. Very recently Diodati et al., [1991] analyzed emission from the Italian volcano Stromboli and argued that it is indeed a self-organized critical phenomenon.

Acknowledgment. Supported by the US department of energy under contract DE-AC02-76- CH00016.

REFERENCES

Bak, P., and K. Chen, Fractal Dynamics of earthquakes, in *Fractals and their Applications to Geology,* edited by C. Barton, Geological Society of America, Denver, in press, 1992.

Bak, P., and K. Chen, Self-organized criticality, *Sci. Am. 246(1),* 46-53, 1991.

Bak, P., and C. Tang, Earthquakes as a self-organized critical phenomenon, *J. Geophys. Res. B94,* 15635-15638, 1989.

Bak, P., C. Tang, and K. Wiesenfeld, Self-organized criticality: an explanation of 1/f noise, *Phys. Rev. Lett. 59,* 381-384, 1987.

Bak, P., C. Tang, and K. Wiesenfeld, Self-organized criticality, *Phys. Rev. A 38,* 364-372, 1988a.

Bak, P., C. Tang, and K. Wiesenfeld, in *Cooperative Dynamics in Complex Systems,* edited by H. Takayama, Springer, Tokyo, pp. 274-284, 1988b.

Burridge, R. and L. Knopoff, *Bull. Seismol. Soc. Am., 57,* 341, 1967.

Carlson, J. M. and J. S. Langer, Properties of earthquakes generated by fault dynamics, *Phys. Rev. Lett. 62,* 2632-2635, 1989.

Chen K., P. Bak, and S. P. Obukhov, Self-organized criticality in a crack propagation model of earthquakes, *Phys. Rev. A 43,* 625-630, 1991.

Christensen, K., and Z. Olami, Variation of the Gutenberg-Richter b values and non-trivial temporal correlations in a spring block model for earthquakes, *J. Geophys. Res.,* in press, 1992a.

Christensen, K. and Z. Olami, Scaling, Phase transitions, and Non-universality in a Self-Organized Cellular Automaton Model, *Phys. Rev. A,* in press, 1992b.

Diodati, P., F. Marchesoni, and S. Piazzo, Acoustic emission from volcanic rocks: an example of self organized criticality, *Phys. Rev. Lett., 67,* 2239-2243, 1991.

Feder, H. J. S. and J. Feder, *Phys. Rev. Lett. 66,* 2669-2672, 1991.

Gutenberg, B. and C. F. Richter, *Ann. Geofis. 9,* 1, 1956.

Ito, K. and M. Matsuzaki, Earthquakes as a self-organized critical phenomenon, *J. Geophys. Res. B95,* 6853, 1990.

Johnston, A. C. and S. J. Nava, Recurrence rates and probability estimates for the New Madrid seismic zone, *J. Geophys. Res. B90,* 6737, 1985.

Kagan, Y.Y., Statistics of "characteristic" earthquakes, *Bull. Seismol. Soc. Amer.,* to be published, 1992.

Kagan, Y. Y. and D. Jackson, Long-term earthquake clustering, *Geophys. J. Int., 104,* 117-133, 1991.

Mandelbrot, B., The variation of certain speculative prices, *The Journal of Business of the University of Chicago, 36,* 394-419, 1963.

Mandelbrot, B., The Fractal Geometry of Nature, W. H. Freeman, San Francisco, 1982.

Olami, Z., H. J. S. Feder, and K. Christensen, Self-organized criticality in a non-conservative cellular automaton modeling earthquakes, *Phys. Rev. Lett. 68,* 1244-1247, 1992.

Raup, M. D., Biological extinction in earth history, *Science, 231,* 1528-1532, 1986.

Scholz, C. H., Earthquakes and faulting: self-organized critical phenomena with a characteristic dimension, in *Spontaneous Formation of Space-time Structure,* edited by T. Riste and D. Sherington, Plenum, New York and London, 1991.

Sornette, A. and D. Sornette, Self-organized criticality and earthquakes, *Europhys. Lett. 9,* 192-195, 1989

Vere-Jones, D., *Pageoph. 114,* 711, 1976; *Math. Geol. 9,* 455, 1977.

Wiesenfeld, K., P. Bak, and C. Tang, A physicist's sandbox, *J. Stat. Phys. 54,* 1441-1451, 1989.

Xu, H-J., B. Bergersen, and K. Chen, A new crack propagation model of earthquakes, preprint, 1992.

Per Bak, Kim Christensen, and Zeev Olami, Brookhaven National Laboratory, Department of Physics, Upton, NY 11973

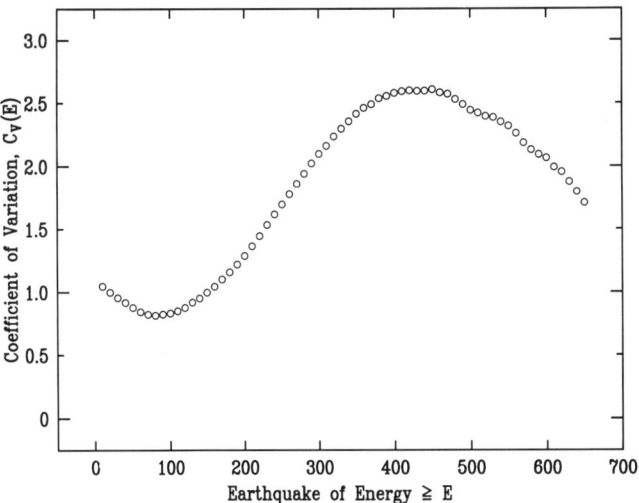

Period-Doubling Bifurcation and Chaotic Phenomena in a Single Degree of Freedom Elastic System with a Two-State Variable Friction Law

NIU ZHIREN AND CHEN DANGMIN

Seismological Bureau of Shaanxi Province, Xi'an, China

A numerical simulation study on the detailed quasi-static behavior of single degree of freedom elastic system with a two-state variable friction law as the system undergoes the Hopf bifurcation is carried out in this paper. The fine structure of the period-doubling bifurcation is studied by investigating phase trajectories, and analyzing power spectra. A sequence of periodic attractors of period 2^n, $n = 0, 1, 2, 3, 4, 5, 6$ is given for this system. Following Feigenbaum, the ratio δ_n of this bifurcation sequence is determined as

$$\delta_1 = 5.75, \quad \delta_2 = 6.03,$$
$$\delta_3 = 3.77, \quad \delta_4 = 3.94.$$

In this paper a chaotic attractor for the system is given in the phase plane, and the chaotic state is entered from quasi-periodic states by a period-doubling bifurcation.

INTRODUCTION

A frictional sliding instability between rock surfaces in the laboratory corresponds at least qualitatively to a shallow depth earthquake instability along an exiting fault (Brace and Byerlee, 1966). In order to examine the predictability of earthquakes, it is worth investigating the nonlinear dynamic behavior of frictional sliding between rock surfaces. Experimental studies on frictional sliding in rock due to Dieterich (1978, 1979, 1981) and Ruina (1980, 1983) have led to a class of constitutive relations for frictional slip resistance which depends on slip rate and slip history as well as on environmental factors, such as pore pressure, rock type, porosity, gouge layer thickness, and effective normal pressure. The general mathematical framework for this class of constitutive relations has the form (Ruina, 1980, 1983)

$$\tau = F(V, \sigma_n, \theta_1, \theta_2, ..., \theta_n)$$
$$\frac{d\theta_i}{dt} = G_i(V, \sigma_n, \theta_1, \theta_2, ..., \theta_n) \quad (1)$$

for $i = 1, 2, ..., n$ where τ is the shear strength, V is the slip velocity, σ_n is the normal stress, and $\theta_1, \theta_2, ..., \theta_n$ are a set of phenomenological parameters called state variables. In slip motion at a fixed slip rate and normal stress, the state variables evolve towards steady state values θ_i^{**} satisfying $G_i(V, \sigma_n, \theta_1^{**}, \theta_2^{**}, ..., \theta_n^{**}) = 0$, such that the shear strength evolves towards a steady state value $\tau^{**}(V)$ corresponding to a fixed speed and normal stress. A fairly comprehensive nonlinear analysis of quasi-static slip motion of a driven spring-block system (Figure 1) and its possible instabilities have been given by Gu et al. (1984), Rice and Gu (1983) and Blanpied et al. (1984) for specific one and two-state variable laws [as special cases of the class (1)]. The one state variable law was proposed by Ruina (1980, 1983), as an approximation to a law proposed by Dieterich (1979, 1981), and the two-state variable law is of similar form but provide a closer fit to the observed relaxations in Ruina's experiments.

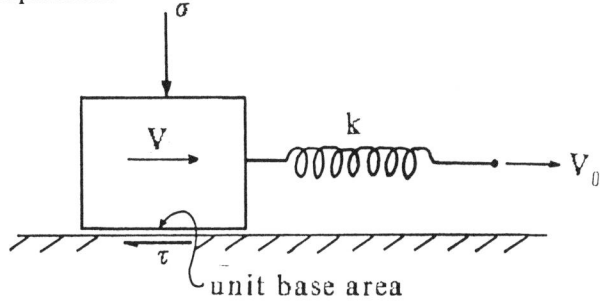

Fig. 1. A single degree of freedom spring-block system. A slider of unit base area is loaded by the driver. σ is normal stress and τ is frictional stress. k is spring stiffness, V_0 and V are velocities of the load point and the block, respectively.

The quasi-state slip motion of a driven spring-block system (Figure 1) with one state variable friction law can be described by an autonomous system of first order ordinary differential equations with two independent variables (Gu et al., 1984). It is well known that period-doubling or chaotic transitions cannot be observed in such a system with two independent variables (Hao, 1989). In this paper, we investigate the detailed dynamical behavior of the quasi-static slip motion of a driven spring-block system with a two-state variable friction law, as this system undergoes the Hopf bifurcation, analyze the fine structure of the period-doubling bifurcations, and elucidate the route to chaos in such a system.

SINGLE DEGREE OF FREEDOM FRICTION ELASTIC SYSTEM WITH TWO STATE VARIABLE FRICTION LAW

A single degree of freedom spring-block system is shown in Figure 1 where k denotes the spring stiffness, while τ and σ are the friction resistance and the normal stress acting on the block, respectively. As the load point is displaced with a rate $V_0 > 0$, the slider block moves with a rate $V > 0$. For quasi-static response of the spring-slider system, the equation of motion is

$$\frac{d\tau}{dt} = k(V_0 - V) \quad . \tag{2}$$

A specific form of the general slip rate and slip rate history dependent law (1) is the two-state law

$$\tau = \tau_* + A \ln(V/V_*) + B_1 \theta_1 + b_2 \theta_* \quad , \tag{3}$$

$$\frac{d\theta_1}{dt} = -\frac{V}{L_1}[\theta_1 + \ln(V/V_*)] \tag{4}$$

$$\frac{d\theta_2}{dt} = -\frac{V}{L_2}[\theta_2 + \ln(V/V_*)] \tag{5}$$

in which θ_1 and θ_2 are state variables; L_1 and L_2 are characteristic displacements; V^* is a reference velocity, for which the steady state stress of friction is τ_*; and A, B_1 and B_2 are constitutive parameters. This law was proposed by Ruina (1980, 1983) in order to fit results over a wide range of slip rates, of approximately 0.01 to 1 μm s^{-1}, with polished quartzite surfaces. As Gu et al. (1984) comment, the same form seems to describe qualitatively experiments with various gouge layers (Dieterich, 1981), except that the L's can be much larger, e.g. of the order of 100 μm. The two-state variable law is schematically illustrated in Figure 2.

For convenience in later discussion, following Gu et al. (1984), the following dimensionless variables

$$f = \frac{\tau - \tau_*}{A}, \quad \varphi = \ln\frac{V}{V_*}, \quad \Theta_1 = \frac{\theta_1}{A}, \quad \Theta_2 = \frac{\theta_2}{A}, \quad T = \frac{V_*}{L_1}t,$$

and dimensionless parameters

$$\beta_1 = \frac{B_1}{A}, \quad \beta_2 = \frac{B_2}{A}, \quad \rho = \frac{L_1}{L_2}, \quad K = k\frac{L - L_1}{A}, \quad \varphi_0 = \ln\frac{V_0}{V_*},$$

are introduced. The set of equations (2)–(5), therefore, are rewritten as

$$\frac{d\Theta_1}{dT} = e^\varphi(\Theta_1 + \beta_1 \varphi), \tag{6}$$

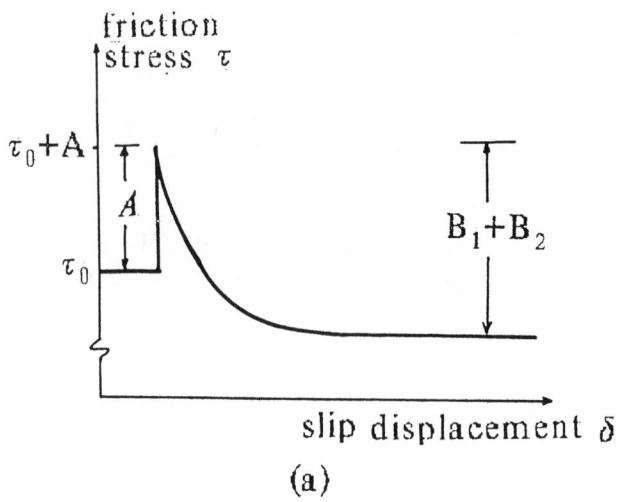

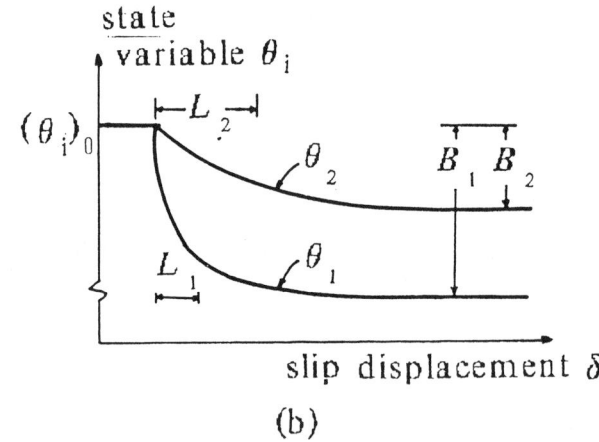

Fig. 2. Response to step change in slip rate for two-state variable friction law of equations (3)–(5). (a) Stress response as the slip rate jumps from V_0 to eV_0; the stress jumps to a peak $\tau_0 + A$ from original steady stress τ_0, then decays exponentially with ongoing slip, with distinct decay slip distances L_1 and L_2, to the new steady state stress $\tau_0 - (B_1 + B_2 - A)$. (b) Decay of the state variables θ_1, θ_2 to new steady state levels.

$$\frac{d\Theta_2}{dT} = -\rho e^\varphi(\Theta_2 + \beta_2 \varphi), \tag{7}$$

$$\frac{df}{dT} = -K(e^\varphi - e^{\varphi_0}), \tag{8}$$

$$f = \Theta_1 + \Theta_2 + \varphi \quad . \tag{9}$$

This is a set of first order autonomous equations with three independent variables Θ_1, Θ_2, and f. Bifurcation and chaos in this friction system will be investigated in the next section.

PERIOD-DOUBLING BIFURCATIONS LEADING TO CHAOS

A simple system may exhibit quite complex dynamical behavior. The simplest example is the logistic map (May, 1976)

$$X_{n+1} = 1 - \mu X_n^2, \quad -1 < X_n < 1, \quad 0 < \mu < 2 \ . \quad (10)$$

The iterations of this relation have a remarkable range of dynamic behavior depending upon the value of μ. For $0 < \mu < 0.75$ the equation iterates to the stable fixed point $X^* = \left(\sqrt{1 + 4\mu} - 1\right)/2\mu$. At $\mu = 0.75$ a flip bifurcation occurs and the asymptotic solution has a 2-cycle. For increasing μ, further bifurcations occur and there come consecutively the 4, 8, ..., 2^n, ... cycles, forming a period-doubling bifurcation sequence. In the range $2 > \mu > \mu_\infty = 1.40115...$, windows of chaos and multi-root cycles are found. Feigenbaum (1978) investigated the convergence of μ_n ($n = 1, 2, 3, ...$) to μ_∞, where μ_n is the bifurcation point from a 2^{n-1}-cycle to a 2^n-cycle. He discovered that they converge geometrically, i.e.,

$$\mu_n = \mu_\infty - \frac{\text{const}}{\delta_n} \ . \quad (11)$$

It is remarkable that

$$\delta_\infty = 4.6692... \quad (12)$$

happens to be a "universal" constant for a large class of mappings. The convergence law (11) shows that δ may be estimated by looking at the ratio

$$\delta_n = \frac{\mu_n - \mu_{n-1}}{\mu_{n+1} - \mu_n} \quad (13)$$

for several consecutive values of μ_n.

Let us now study bifurcation and chaotic behavior in the system (6)–(9). This system has been integrated numerically using a fourth order Runge-Kutta scheme. The dynamic evolution of the system has been investigated for different values of the dimensionless stiffness K under zero initial conditions and at fixed $\beta_1 = 1.00$, $\beta_2 = 0.84$, $\rho = 0.048$, $\varphi_0 = 0.19885$. The computed results show that the investigated system can have a complex dissipative structure (Glansdorff and Prigogine, 1971; Nicolis and Prigogine, 1977), i.e., the system can evolve towards a new equilibrium state, a periodic state, or a chaotic state when the parameter K changes. Figures 3, 4 and 5 give orbits in the (f, φ) phase plane for $K = 0.06983$, 0.0685, and 0.0683, respectively. These orbits are recorded only after the system reaches a statistically steady state. It is observed that Figures 3, 4, and 5 give a limit cycle, a chaotic orbit, and an unstable orbit, respectively.

Further information on system stability is obtained from a global bifurcation diagram which illustrates the qualitative variations of system behavior with respect to model parameters. The global bifurcation of orbits is analyzed for different K. The analyzed results are summarized in Table 1.

For decreasing K from 0.0699 to 0.0685, the bifurcations occur and a sequence of period-2^n bifurcations, $n = 0, 1, 2, 3, 4, 5, 6$, are observed in the phase plane. Figures 6, 7 and 8 give period-2^n orbits in the (f, φ) phase plane, n = 1, 2 and 4, respectively. At $k = 0.0685$, deterministic chaos occurs (see Figure 4).

In order to arrive at a deeper understanding of the dynamics, power spectrum analysis of the dynamics has been carried out here. The transition to chaos from the quasi-periodic regime is recognized by means of the power spectra. Figures 9, 10, 11 and 12 give the power spectra of period-1, period-2, period-16, and chaotic orbits, respectively.

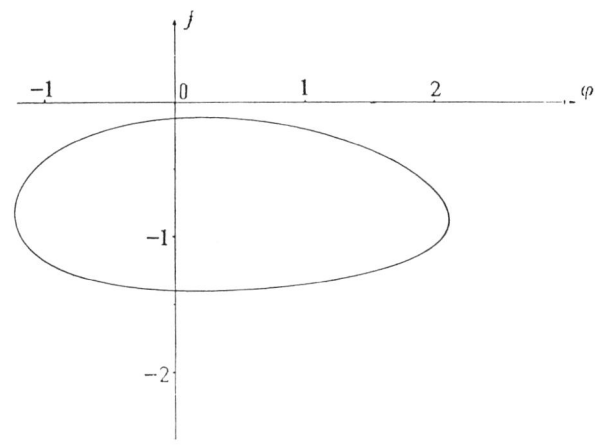

Fig. 3. Limit cycle, a periodic state, of the system (6)–(9). $K = 0.06983$.

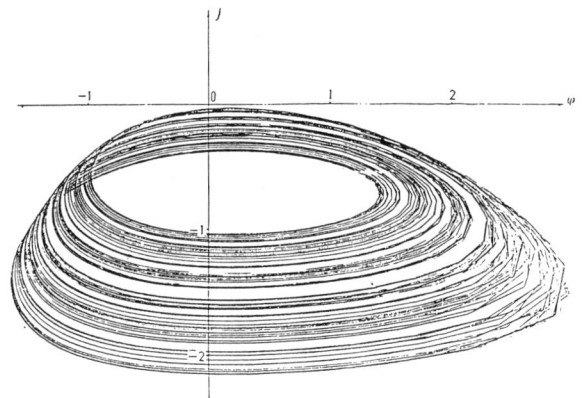

Fig. 4. Chaotic evolution, a chaotic state, of the system (6)–(9). $K = 0.0685$.

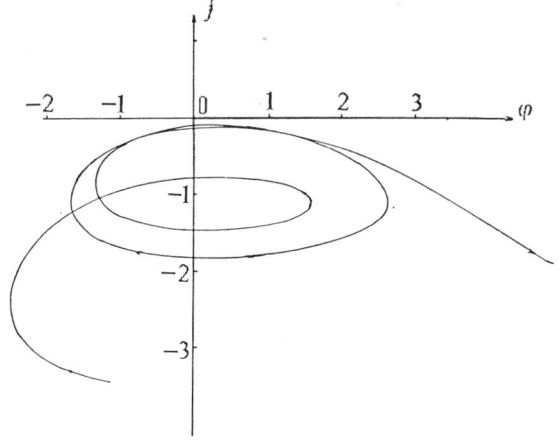

Fig. 5. Unstable orbit in the (f, φ) plane of the system (6)–(9). $K = 0.0683$.

TABLE 1. Period number of asymptotically stable orbits of the system (6) ~ (9) for different non-dimensional stiffness K.

dimensionless stiffness. K	period number	dimensionless stiffness. K	period number
0.06990	1	0.0686375	16
0.06983	1	0.0686370	16
0.06967	1	0.0686350	16
0.06965	2	0.0686330	16
0.06950	2	0.0686320	16
0.06882	2	0.0686315	32
0.06880	4	0.0686310	32
0.06870	4	0.0686300	32
0.0686625	4	0.0686297	32
0.0686620	8	0.0686295	64
0.0686400	8	0.0686290	64
0.0686380	8	0.0686280	64

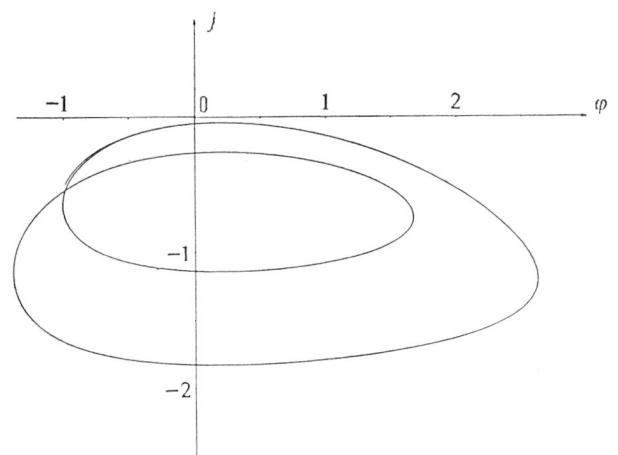

Fig. 6. Period-two limit cycle of the system (6)–(9). $K = 0.06882$.

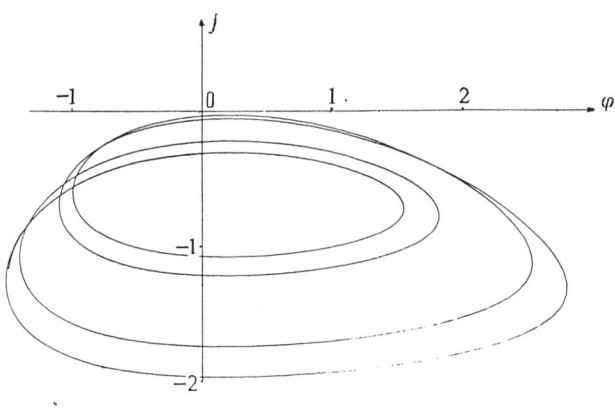

Fig. 7. Period-four limit cycle of the system (6)–(9). $K = 0.06866\text{-}25$.

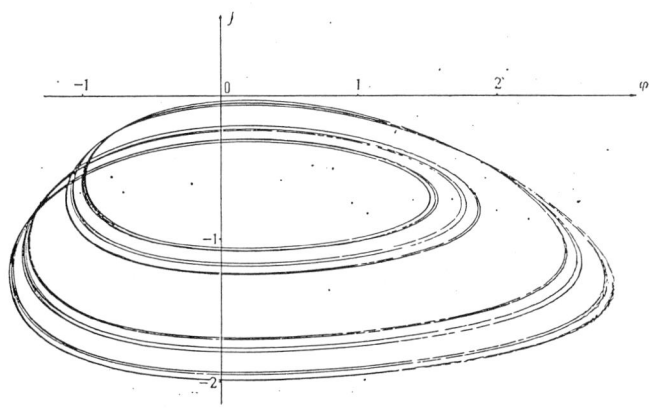

Fig. 8. Period-sixteen limit cycle of the system (6)–(9). $K = 0.068633$.

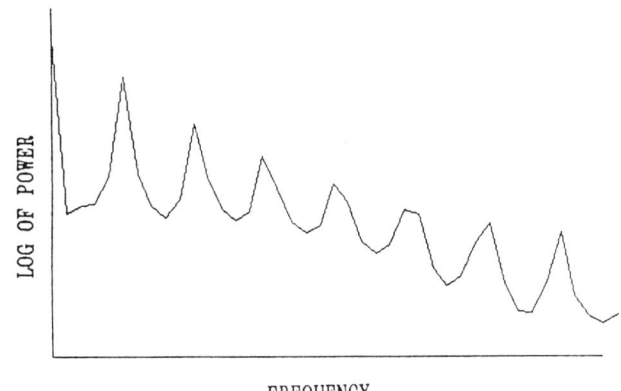

Fig. 9. Power spectrum of the period-1 orbit. $K = 0.06980$.

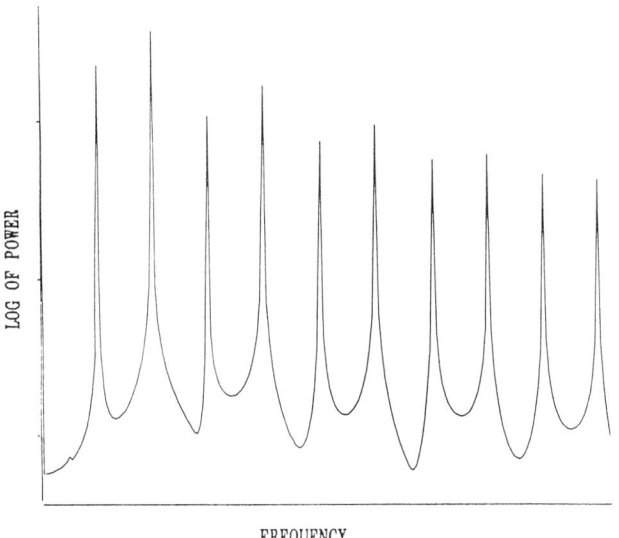

Fig. 10. Power spectrum of the period-2 orbit. $K = 0.06883$.

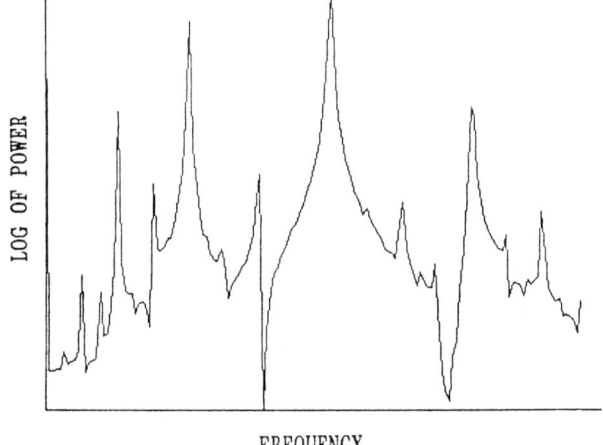

Fig. 11. Power spectrum of the period-16 orbit. $K = 0.068633$.

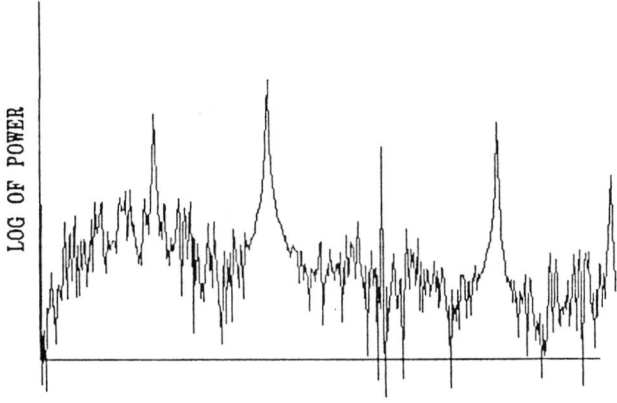

Fig. 12. Power spectrum of the chaotic orbit. $K = 0.06850$.

Following Feigenbaum (1978, 1979, 1983), the ratio δ_n (13) of the investigated system is determined as

$$\delta_1 = 5.75, \quad \delta_2 = 6.03, \quad \delta_3 = 3.77, \quad \delta_4 = 3.94.$$

These values are close to the Feigenbaum number $\delta = 4.6692....$ However, they do not show clear sign of convergence. It is possible that the manner of convergence for the bifurcation sequence to chaos for the friction system is different from that for the logistic map.

Conclusions

(1) The evolution of a single degree of freedom friction elastic system with a two-state variable friction law can have a complex dissipative structure. Equilibrium points, limit cycles and chaotic orbits are all found for different values of the dimensionless stiffness K.

(2) A sequence of period-doubling bifurcations in the system is given and the ratio δ_n of this bifurcation sequence is determined as

$$\delta_1 = 5.75, \quad \delta_2 = 6.03, \quad \delta_3 = 3.77, \quad \delta_4 = 3.94 \ .$$

It is possible that the convergence manner of the bifurcation points to the chaos point for the friction system is different from that for the logistic map.

(3) This study elucidates the route to chaos in the investigated system and gives a chaotic orbit in the (f, φ) phase plane.

(4) Our results provide evidence that the stick-slip evolution of a fault system has complex dynamic behavior. It seems that the stick-slip evolution of a fault system is inherently unpredictable. This study indicates that earthquake precursors and earthquake process are complex.

(5) The present study gives a quasi-static analysis. It is worth investigating the effects of inertia in the friction system at lower stiffness.

References

Blanpied, M., T. Tullis and J. Weeks, Stability and behavior of frictional sliding with a two-state variable constitutive law (abstract), *Eos Trans. AGU*, *65*, 1077, 1984.

Brace, W.F., and J.D. Byerlee, Stick-slip as a mechanism for earthquakes, *Science*, *153*, 990-992, 1966.

Dieterich, J.H., Time-dependent friction and the mechanics of stick-slip, *Pure Appl. Geophys.*, *116*, 790-806, 1978.

Dieterich, J.H., Modeling of rock friction, 1, Experimental results and constitutive equations, *J. Geophys. Res.*, *84*, 2161-2168, 1979.

Dieterich, J.H., Constitutive properties of faults with simulated gouge, in *Mechanical Behavior of Crystal Rocks, Geophys. Monogr. Ser.*, Vol. 24, edited by N.L. Carter, M. Friedman, J.M. Logan, and D.W. Stearns, pp. 103-120, AGU, Washington, D.C., 1981.

Feigenbaum, M.J., Quantitative universality for a class of nonlinear transformations, *J. Stat. Phys.*, *19*, 25, 1978.

Feigenbaum, M.J., The universal metric properties of nonlinear transformations, *J. Stat. Phys.*, *21*, 669, 1979.

Feigenbaum, M.J., Universal behavior in nonlinear systems, *Physica*, *7D*, 16, 1983.

Glansdorff, P., and I. Prigogine, *Thermodynamics of Structure, Stability and Fluctuations*, Wiley, London, 1971.

Gu, J.-C., J.R. Rice, A.L. Ruina, and S.T. Tse, Slip motion and stability of a single degree of freedom elastic system with rate and state dependent friction, *J. Mech. Phys. Solids*, *32*, 167-196, 1984.

Hao, Bai-Lin, *Elementary Symbolic Dynamics and Chaos in Dissipative Systems*, WSPC, Singapore, 1989.

May, R.M., Simple mathematical models with very complicated dynamics, *Nature*, *261*, 459-467, 1976.

Nicolis, G., and I. Prigogine, *Self-Organization in Nonequilibrium Systems*, Wiley, New York, 1977.

Rice, J.R., and J.-C. Gu, Earthquake aftereffects and triggered seismic phenomena, *Pure Appl., Geophys.*, *121*, 187-219, 1983.

Ruina, A.L., Friction Laws and Instabilities: A Quasistatic Analysis of Some Dry Frictional Behavior, Ph.D. thesis, Brown Univ., Providence, R.I., 1980.

Ruina, A.L., Slip instability and state variable friction laws, *J. Geophys. Res.*, *88*, 10359-10370, 1983.

Niu Zhiren and Chen Dangmin, Seismological Bureau of Shaanxi Province, Xi'an, China.

Nonlinear Dynamic Modeling of Earthquake Prediction

YAOLIN SHI

Graduate School, Academia Sinica, Beijing 100039, China

LUMIN GENG   GUOMIN ZHANG

Center for Analysis and Prediction, State Seismological Bureau, Beijing 100036, China

Earthquake prediction has been a major project in Chinese seismology for twenty-five years. Understanding of the mechanism of earthquake precursors is important for making further advances. In this paper, we use a network of block-spring-dashpot models to simulate continental seismicity of interrelated fault zones. The results provide insights to the process of seismogenesis. Events produced in the modeling show remarkable similarity to realistic seismicity in their temporal, spatial and magnitude distribution. The system behaves noisily although the model is deterministic. Stress variation calculated in the model helps in understanding the behavior of the system. Predictability of large events is investigated. It is suggested that a comprehensive method, combining statistical analysis of seismicity and observation of stress-related precursors, may provide a better chance for earthquake forecasting.

INTRODUCTION

Earthquake prediction has been an important scientific project in China since the Xingtai earthquake of 1966, which caused serious damage to life and property in a region only three hundred kilometers from Beijing. Early immature optimism has faded after twenty-five years of practice mixed with astonishment and frustration. One might imagine that earthquake precursors would occur around a forthcoming earthquake epicenter before a major shock, however, no such ideal pattern exists. Anomalies may occur without a major earthquake having occurred, and all precursors do not appear before a major earthquake. Observed anomalies before large earthquakes are not limited to the future epicenter area, instead they are scattered over a broader region [e.g., Ma et al., 1990]. Facing this complexity, it has been agreed by most seismologists that the understanding of the physical mechanism of earthquake precursors has become a critical issue, if further advances in earthquake prediction are to be expected.

A complete analysis of the genesis of an earthquake is still beyond our present knowledge of earthquake source physics and our capacity of computation. However, simplified models have been intensively studied in order to understand the essence of earthquake occurrence. A block-spring model was introduced by Burridge and Knopoff [1967]. A lot of studies have been executed along this direction. Experimental work on friction has been investigated to improve our understanding of friction laws [e.g., Byerlee 1970, 1978]. Cao and Aki [1986] used a rate- and state-dependent friction law to simulate seismicity. Nussbaun and Ruina [1987] examined the complex behavior of a simple two degree-of-freedom system, Gabrielov et al. [1986] used a multilayer block model to study lithosphere dynamics. Carlson and Langer [1989] and Langer and Tang [1991] used fault models with stick-slip velocity weakening friction nonlinearity to model properties of earthquakes. Bak and Tang [1988] applied a cellular automation model of self-organized critical phenomena to earthquake study, Brown et al. [1991] combined the concept of a block-spring model with cellular automation. Although extensive study based on state dependent friction laws have been carried out, most of the above studies were limited to the fracture of one single fault. To study a fault system like those observed in China, Zhu and Shi [1991] used a group of parallel slide elements in series to model a system with several faults. In this paper, we will use a network of block-spring-dashpot models to model earthquake occurrence in several fault zones with interaction, which are common in continental seismic zones. Characteristics of the synthetic events will be compared with real earthquakes. Predictability of large events will be of special interest.

NONLINEAR MODEL

The basic element in our model consists of a Maxwell body (a spring and a dashpot in series) and a rigid sliding block.

Sliding takes place when stress in the sliding block exceeds its static strength, and sliding stops as soon as stress drops to its dynamic strength. Each element represents a segment of a fault, and a group of parallel elements represents the entire fault. A network of elements, consisting of several groups of parallel elements connected by coupling elements, represents a system of several faults with interaction (Figure 1). A break of an element simulates an earthquake, and simultaneous breaks of several neighboring elements simulates a large earthquake.

The system can be described by a group of linear differential equations

$$\dot{\sigma}(t) = A\sigma(t) + B\dot{\varepsilon}(t) \qquad (1)$$

where σ is the stress vector, $\dot{\varepsilon}$ is the strain rate vector, and matrix A and B can be obtained from the configuration of the network and mechanical properties of the elements, based on the principles that forces at each node are in balance and the total strain rate of the system is equal to the summation of the strain rate of elements in series. The derivation of the differential equation, similar to that of an electric circuit, consisting of resistance and capacities, has been well studied in electric engineering. If the total strain rate enforced on the system is known, being chosen to be a constant strain rate in this study, the solution is obtained

$$\sigma(t) = \Phi(t)\sigma(0) + \int_0^t \Phi(t-\tau)B\dot{\varepsilon}(\tau)d\tau \qquad (2)$$

where

$$\Phi(t) = exp(At) = \sum_{k=0}^{\infty} \frac{(At)^k}{k!} \qquad (3)$$

which can be solved numerically. Variation of stress with time in each element therefore can be calculated.

When stress in an element exceeds its static strength, the element will break, and its stress will drop to its dynamic strength. Stress drop in one element will cause stress adjustment in the whole system. The dashpots do not affect the instantaneous change of stress. Changes in stresses can be described by a group of algebraic equations related to the elastic properties of elements.

Figure 2 shows the readjustment of stresses in a network of 6 columns and 8 rows, corresponding to six parallel faults each consisting of eight segments. A break of an element results in an increase of stress in neighboring elements on the same fault, the magnitude of stress increase depending on the distance between the element and the broken element. Those closest to the broken element have the greatest increase in stress. In elements of all the other faults, stresses usually decrease. Maximum reduction of stress in each fault occurs at elements in the same row as the broken element, and the magnitude also decays with their distance from the broken element. On each fault, magnitude of stress reduction usually decreases away from the element with maximum change on that fault, with greater gradient on faults closer to the broken fault. It is noted that in faults in the immediate neighbor to the broken faults, stresses in some elements may increase slightly, as a result of the combined effect of stress drop in the broken element and stress rise in its neighbors on the same fault.

Adjustment of stresses strongly depend on the coupling elements. If the elastic modula of coupling elements approaches infinity, the network becomes a simple case of several parallel elements in series. Stress drop in one element will lead to stress increase in other elements on the same fault, and the relative increase is equal to each other without correlation of their distance to the broken element. Similarly, all elements on other faults will have the same amount of percentage decrease of stress. This extreme case has been investigated by Zhu and Shi [1991]. Reducing the elastic modula of all coupling elements to zero will produce another extreme. It becomes a very simple case that elements in series form a chain with no relation to other chains. This kind of model has no practical use for earthquake study.

When stress in an element exceeds its static strength, a break of an element occurs. Energy released by the event is $E_n = \sigma^2 / 2E$. As we have mentioned, a break of one element will lead to stress increases in neighboring elements. Therefore, stresses in those elements may sometimes also exceed their strength and a break of several neighboring elements may occur. Although breaking of these elements is sequential, just like the propagation of an earthquake fault at

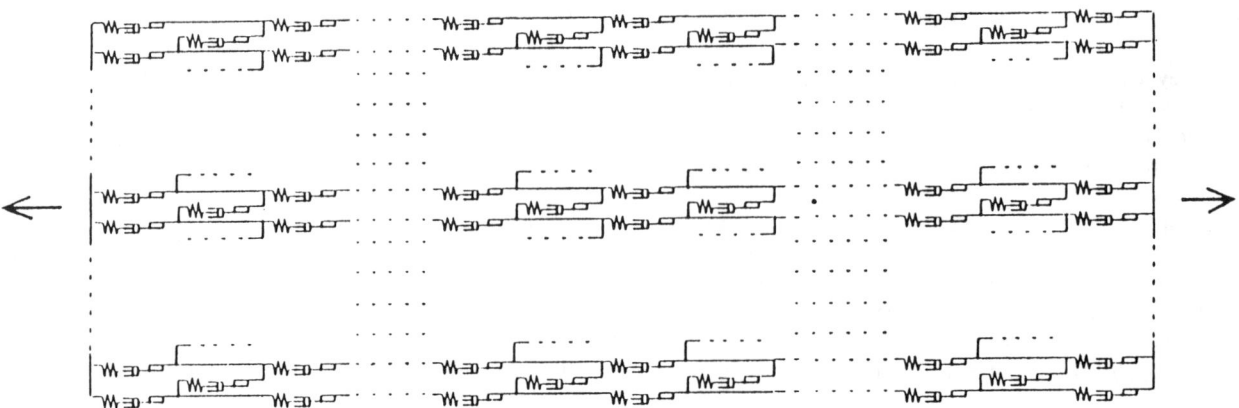

Figure 1. Network of our nonlinear dynamic model. Basic elements in a column represent a fault, and coupling elements connecting different rows simulate the interaction among faults and fault segments. See text for details.

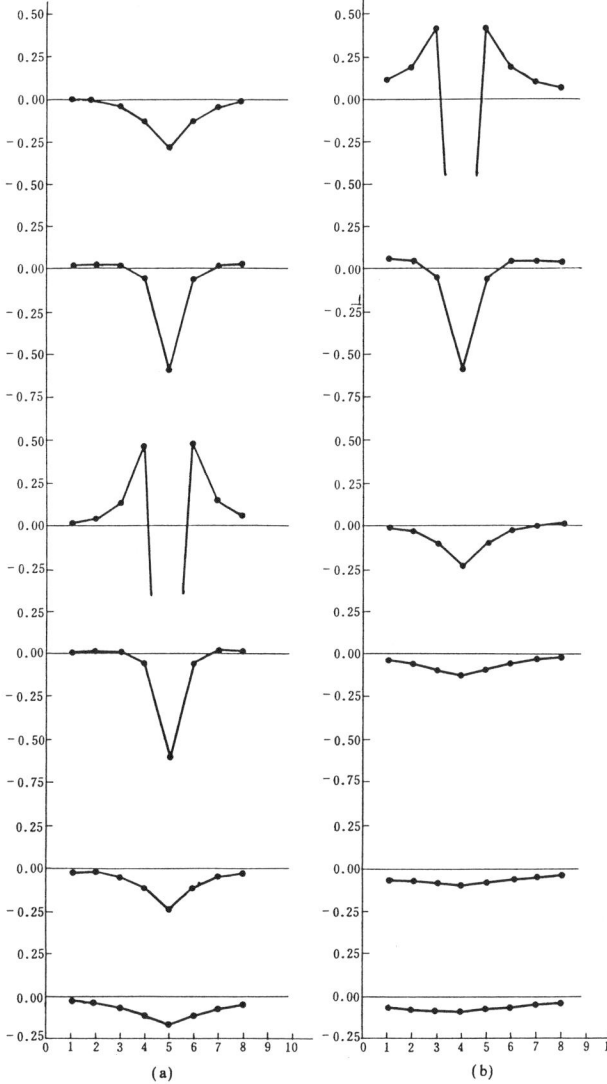

Figure 2. Adjustment of stress after one element is broken in a system consists of 6X8 elements. (a) Element (3,4) is broken. (b) Element (4,1) is broken. In both cases, stress drop of the broken element is 2.0.

finite speed, the process is so fast that it can be regarded as being simultaneously broken in our modeling and is defined as a large earthquake. The total released energy is the summation of all broken elements. Magnitude of these synthetic events is defined as

$$M = A + \log E_n \quad (4)$$

where A and B are constants, in this paper it is chosen to be 3.5 and 1 respectively, to make the smallest event $M = 0$.

Starting with an arbitrary initial condition (usually zero stress for all elements in this paper) we solve the differential equations in order to calculate stress variation with time, then, solving the algebraic equations to calculate the stress adjustment, we take the updated stresses as the new initial condition in order to continue the calculation of stresses. Repetition of the process produces a synthetic earthquake catalog.

RESULTS

Several models with different combinations of rows and columns as well as different mechanical properties have been calculated, the results of a model with 3 columns and 12 rows (3 faults each has 12 segments) are given here. Values of the elastic modulus and viscosity of all elements are chosen to be the same, but static strength of every element is randomly assigned within a range of factor 2. Dynamic strength of an element is chosen to be 0.8 of its static strength [Byerlee, 1970]. This result is representative. Differences of various models will be briefly addressed when necessary.

Figure 3 shows the M−t plot of the relative large synthetic events. Figure 3b show the events which occurred in the entire system, and Figures 3c, 3d and 3e show events in the three faults respectively. It is noted that seismicity is episodic, the active period and quiescent period appearing one after another. For different faults, their characteristics of seismicity are different. Largest earthquakes always occur on fault 3, small earthquakes (M < 1.0) are more frequent on fault 1. These characteristics depend on strength distribution on each fault. Comparing the M−t plots of the three faults, it is noted that quiescence in one fault is usually accompanied by activity in another fault. This is similar to the observed migration of seismicity among different seismic zones in China. The spatial transition of events will be discussed later.

To see the temporal and spatial changes more clearly, Figure 4 show the details of synthetic events in a period of relative time from 5178 to 5578. Figure 4a shows the M−t plot. Figure 4b shows the averaged stress in the system. Figure 4c shows the elastic energy density stored in the system. These two figures show the seismic episodity more clearly. The downslope part represents the active episode, and the upslope part represents the quiescence episode. The two figures are calculated directly in our modeling. These kind of figures can not be obtained directly for real earthquakes, however, they can be estimated indirectly. The strain rate can be estimated from the energy release curve like Figure 4e, then $\dot{\varepsilon}t - \sum E_n$ gives the energy density curve, and its square root is proportional to the averaged stress. Figure 4d shows the accumulated energy release curves for the entire system and for each fault. Although, in a long run, the system shows behavior similar to the slip−predictable model [Shimazaki and Nakato, 1980], it is neither a time−predictable nor a slip−predictable model in a short time period or in a single fault. This result remains valid no matter how the configuration and mechanical property change.

One of the important methods applied in practical earthquake prediction is through the study on spatial and temporal abnormal seismicity, such as seismic gap, seismic banding, quiescence and unusually active seismicity, b value, etc.. It would be interesting to see if these methods can be applied to the prediction of our synthetic events.

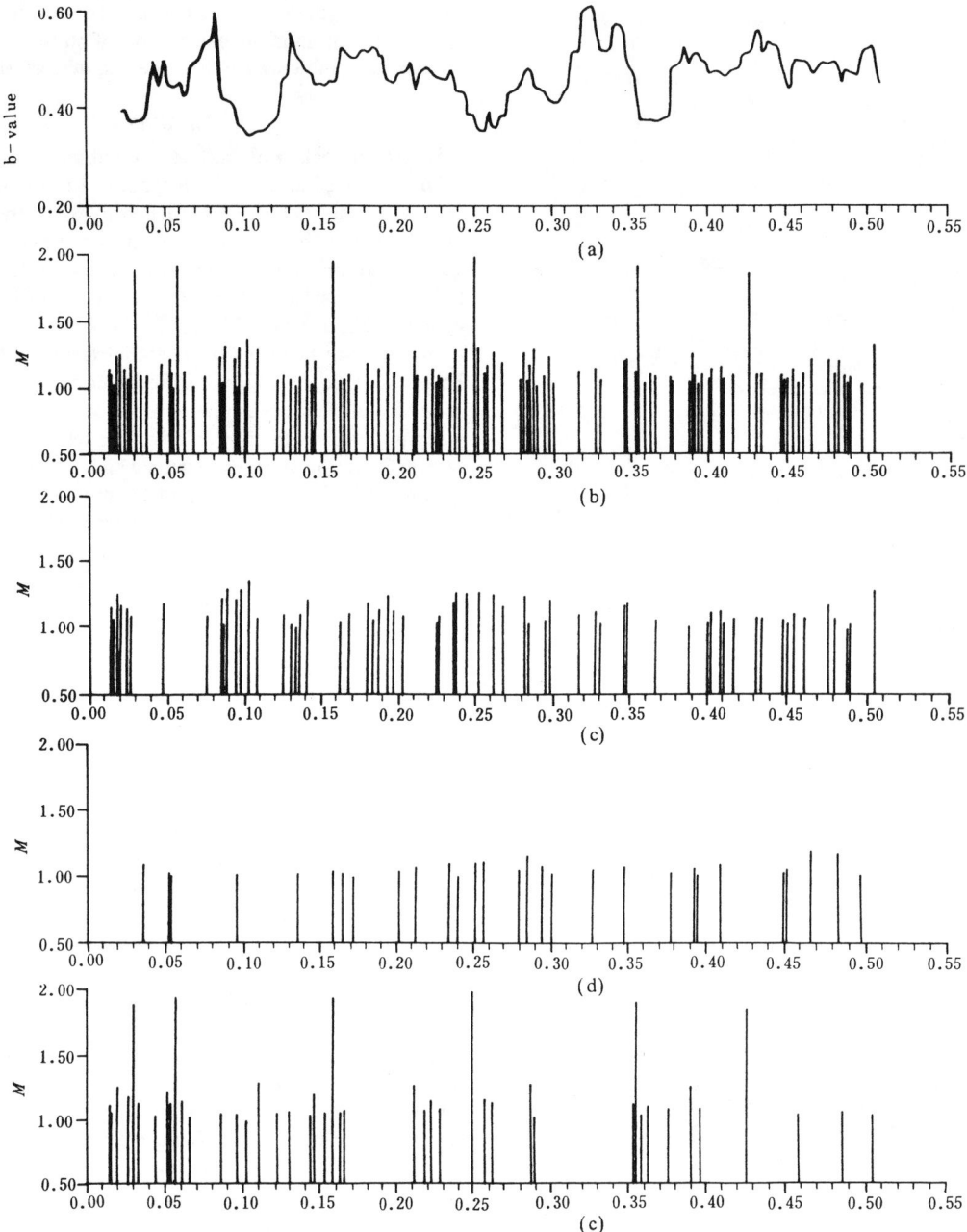

Figure 3. Synthetic events produced by the system. (a) Variation of b value with time, b value is obtained by least square method with a sliding time window. (b) Magnitude versus time for the entire system. (c), (d) and (e) M–t plots for faults 1, 2 and 3 respectively.

Energy release of the synthetic events approximately follows the log N = a − b M relation (Figure 5), but usually both the largest events and the smallest events occur in short of this distribution. There are no clear cutoff magnitudes for such deviation. This is different from results of Carlson [1991], who finds a clear cutoff magnitude, smaller events obey the Gutenberg–Richter relation whereas greater events are in excess of the distribution. The b value depends on the strength distribution of the elements. The value of b tends to be smaller if the differences of the strength of elements become smaller. This is in agreement with laboratory observation that b value is a measure of the heterogeneity of rocks [Mogi, 1962].

The variation of the b value with time is also given in Figure 3a. It is also similar to that observed in real seismicity study

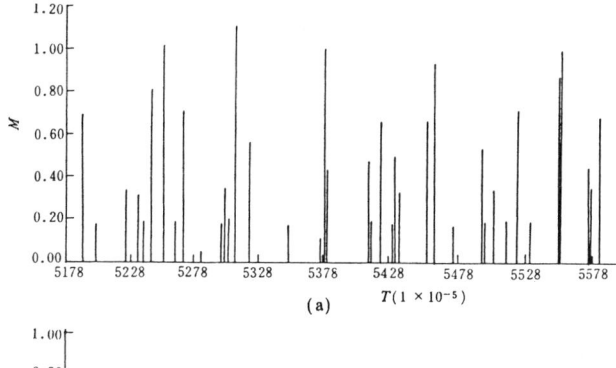
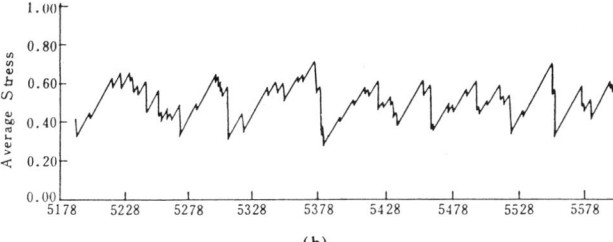
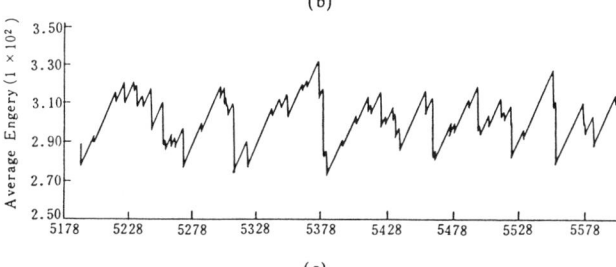
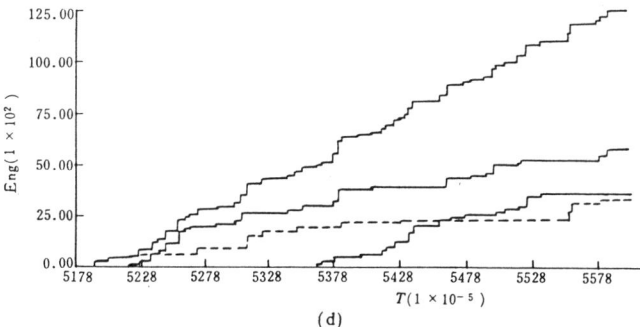

Figure 4. Synthetic events in a period of relative time 5718 to 5578. (a) The M−t plot. (b) Averaged stress in the system. (c) Elastic energy density stored in the system. (d) Accumulated energy release curves for the entire system and for each fault.

[e.g., Figure 3 of Shi and Bolt, 1982]. Decrease of b value was thought to be related to the increase of stress level [Scholz, 1968], and it has been widely used as a earthquake precursor in China [Ma et al., 1990]. Comparing Figures 3a and 3b, a weak correlation between decrease of b value and occurrence of a large event may exist, although the statistical significance of the correlation is not yet clear.

To investigate the spatial migration of the synthetic events, Tables 1 and 2 show the transfer probability matrix of earthquake occurrence in the system. For each fault, after an event occurs, the frequency of occurrence of events in all the three faults are counted to obtain the transfer probability. For example, in Table 1, after the occurrence of an event with magnitude greater than 1.0, the probability of occurrence of the next event is 0.57, 0.15, and 0.28 for Faults 1, 2 and 3 respectively. We further examine if this probability is relatively stable during the entire time. With a sliding window in time, fluctuation of the values of the matrix are found within about 10%. This result can be seen by comparing Tables 1 and 2, their values are for the first half and the second half of the entire time period of modeling. Although the values of the components of the transfer probability matrix depend on the configuration and mechanical property of the network, the time−invariant nature remains valid once the network is chosen. This nature has an important implication for the statistical prediction of the location of forthcoming large earthquakes.

Figure 6 shows the distribution of recurrence time of large events at two different time windows. Figures 6a and 6c are the accumulated frequency distribution for events greater than magnitude 1.0 during the first half period and the last time period respectively. The theoretical Poisson distribution is also drawn for comparison. Figures 6c and 6d show the logarithmic distribution of event frequency for the two time windows respectively. Again, the distributions do not show significant change with time.

Nishenko and Buland [1987] studied the recurrence time of 50 earthquakes and concluded that the normalized recurrence time obeys a lognormal distribution, i.e., the next earthquake has a probability of 0.683 to occur in the time interval

$$T_{ave} e^{\nu - \sigma} < T < T_{ave} e^{\nu + \sigma} \quad (5)$$

where $\nu = -0.020$ and $\sigma = 0.205$ in their sample. We get $\nu = -0.40$ and $\sigma = 0.96$ for our synthetic events. Because this distribution is stable with time, it may be used in prediction of the recurrence time of large events.

In our example, the accumulated distribution is not distinctly different from Poisson distribution, a distribution usual for

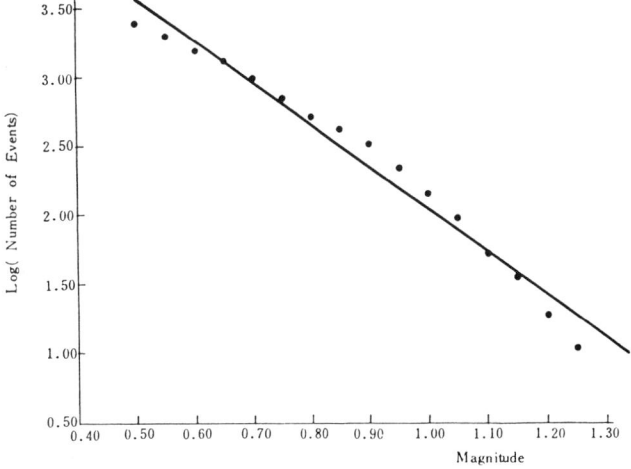

Figure 5. Frequency−magnitude relation for the synthetic events.

Table 1. Transfer probability matrix for the first half period

		Second Event		
		Fault 1	Fault 2	Fault 3
First Event	Fault 1	0.57	0.15	0.28
	Fault 2	0.35	0.35	0.30
	Fault 3	0.30	0.29	0.41

independent events. This is probably a result of a relatively weak coupling chosen for our model, a break in the northern region hardly has any effect on stresses in the southern elements. For different configuration and increased coupling, the difference between the Poisson distribution and the synthetic events becomes more notable.

Many earthquake precursors have been applied to practice in addition to abnormal seismicity, such as stress, deformation, underground water level, Radon content, electric resistivity, geomagnetism, etc.. Most of these anomalies are believed to be related to rock dilatancy under increased stress field. Therefore, we plot the stress variation in the system in order to investigate characteristics of stress related precursors. Locations where elements are under high stresses are labeled by different symbols (Figure 7a). To show the overall state of stress in the system, the total number of locations with anomalies (defined as stress greater than 0.94 of element strength) is also plotted (Figure 7b). Only a small time interval is shown here due to page limitation, but the information is apparent.

It is noted that at low stress levels, only small events occur, while large events are preceded by long lasting high stresses. A large event shown in Figure 7 at relative time 396 is preceded by anomalies started at time 196. Although anomalies appear both in the southern region and the northern region, no large events occur in the southern region. Although the total number of locations with anomalies decreases in the entire system as 11−15−16−15−14−12−10, number of anomalies increases in the northern region as 3−6−7−8−8−9−10. If we denote the intensity of anomaly by a scale of 4, 3 2, 1, and 0.5 for labeled locations " 4 "," 3 "," 2 ","1 ", and " : " respectively, the total intensity (summation of the scaled numbers) increases dramatically in the northern region as 4−9.5−12−15−16.5−21.5−27. Therefore, the stress−related precursors may be a more direct way to predict the occurrence of large events. We noticed, however, both in fault 2 and fault 3, that anomalies appear and evolve on a similar scale, making it difficult to predict the exact location of the forthcoming large event. In this case, statistical analysis of seismicity should be combined with stress−related precursors. A large event of relative time 0 occurs in fault 3 before the event of relative time 396. It is shown in Table 1 that after a large event in fault 3, the probability of occurrence of next large event is 0.40−0.41 in fault 3 and 0.26−0.29 in fault 2. Thus, it is possible to predict that the chance of a large event to occur is 40% higher on fault 3 than on fault 2. Similarly, the recurrence time distribution should be helpful to predict the time of occurrence. Of course, for real earthquakes, high stress may not always lead to detectable anomalies, and disturbance may appear in low stress region, the observed anomalies may not always equate to high stress locations in our simple model, but it is important to illustrate

Table 2. Transfer probability matrix for the second half period

		Second Event		
		Fault 1	Fault 2	Fault 3
First Event	Fault 1	0.58	0.18	0.24
	Fault 2	0.34	0.35	0.31
	Fault 3	0.33	0.26	0.40

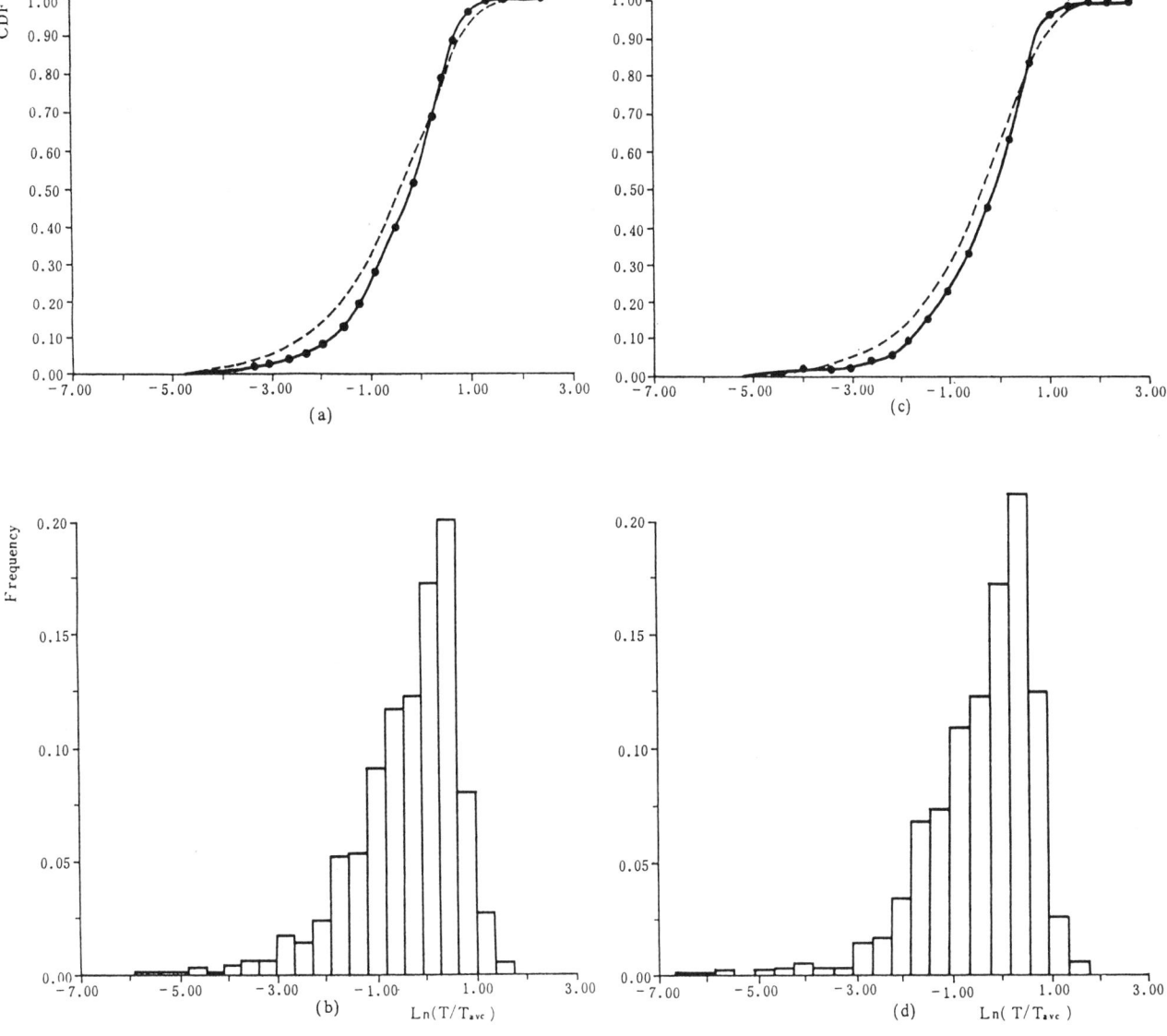

Figure 6. Normalized recurrence time distribution. Frequency is shown in (b) and (d), and accumulated frequency is shown in (a) and (c) for the first and second half time period respectively.

the idea of a comprehensive method combining observation of stress-related precursors and statistical analysis of seismicity.

Analysis of the stress variation also helps to understand migration of earthquakes. It is observed in North China that strong earthquakes migrate from the western seismic zone of Shangsi to the central and eastern seismic zones of Heibei and Bohai in the past several hundred years. Our modeling indicate, that the breaking of a segment along a fault usually increases stresses on the fault and decrease stresses on other faults, this would favor stress accumulation and event occurrence to remain at the same fault, until stress in another fault reaches some critical state to overturn the situation, migration of large events among fault zones.

Conclusion

We proposed a network model of block-spring-dashpot to simulate seismicity in continental seismic zones. Events produced in the modeling show remarkable similarity to realistic seismicity, such as the power law frequency-magnitude relation and the active-quiescence episodes. Stress variation in each element is very complicated and hard to predict, the system may behave chaotically although the model is deterministic. The spatial-temporal distribution of events in the system appears random in a short duration; the overall behavior of the system, such as the recurrence time and the spatial migration of large events, show regularities if the time

NONLINEAR DYNAMIC MODELING

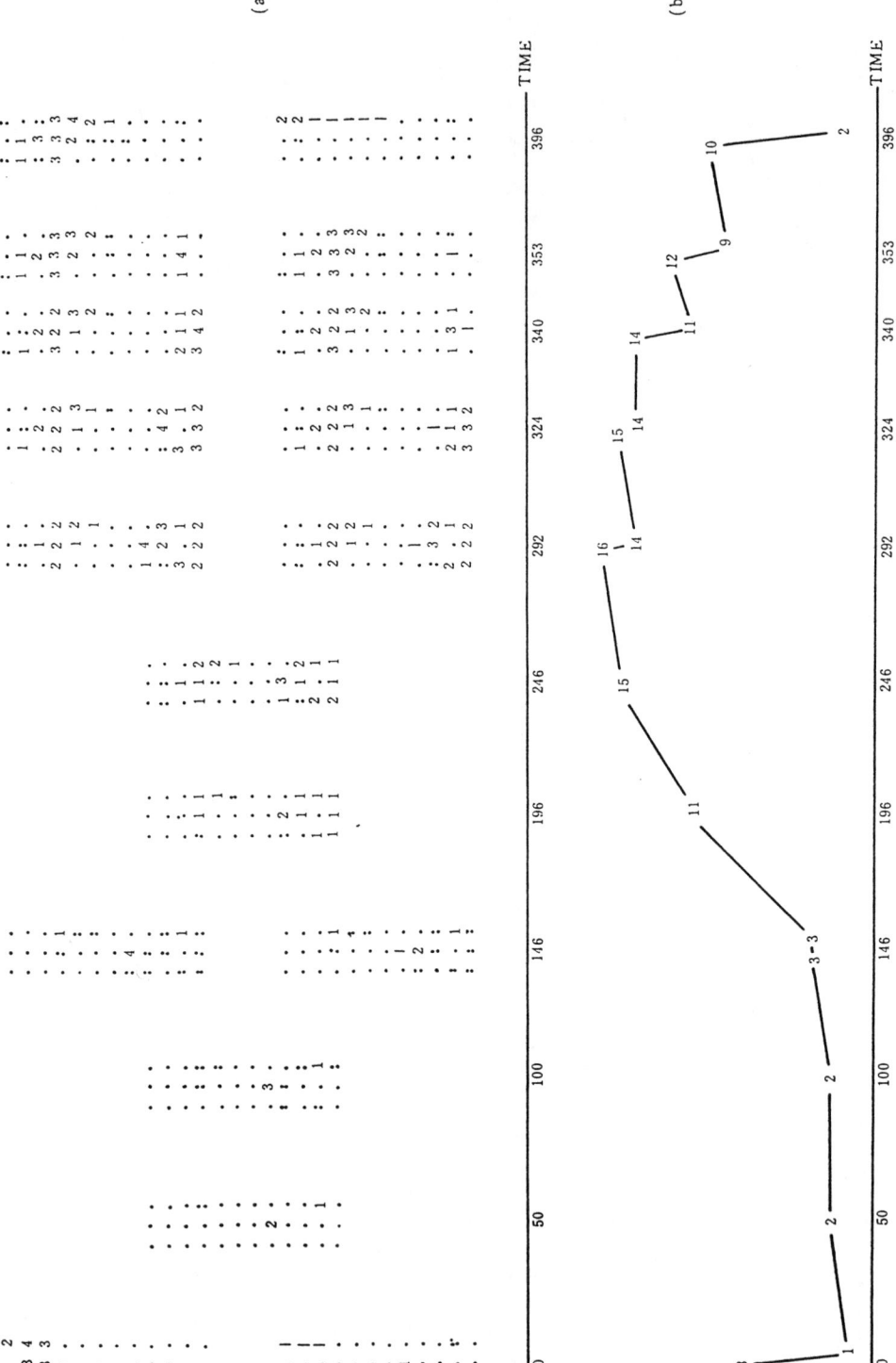

Figure 7. (a) Locations of high stresses and their evolution with time. location of each element is denoted by " . ", locations with stress of 0.92 to 0.94 of its static strength is denoted by " : ", elements with stress 0.94–0.96, 0.96–0.98, 0.98–1.00 and element immediate before its failure are denoted by " 1 ", " 2 ", " 3 ", and " 4 " respectively. If a slide event occurs, stress state before and after it are given in the upper and lower part respectively, the faulted segments are denoted by "|". (b) The total number of locations with stress greater than 0.94 of its strength.

span is long enough. It may have important implications for probabilistic prediction of real earthquakes.

Variation of stresses with time are calculated in our model. Stress adjustment after an earthquake leads to increase of stress in some locations and decrease of stress in some other locations. It provides insight into the process of migration of large earthquakes and into spatial—temporal characteristics of observed stress—related precursors.

It is suggested that statistical analysis of seismicity and observation of stress—related precursors should be incorporated, A comprehensive method provides a better chance to predict the time and location of large earthquakes.

Acknowledgement. We thank G. Shkurkin for proofreading the text. This study is supported by the National Natural Science Foundation of China.

REFERENCES

Bak, P., C. Tang, Earthquake as a selforganized critical phenomenon, *Jour. Geophys. Res.*, 94, 15635–15637, 1989.

Brown, S. R., C. H. Scholz, J. B. Rundle, A simplified spring—block model of earthquakes, *Geophys. Res. Lett.*, 18, 215–218, 1991.

Burridge, R., and L. Knopoff, Model and theoretical seismicity, *Bull. Seismol. Soc. Am.*, 57, 341–371, 1967.

Byerlee. J. D., Static and kinetic friction of granite at high normal stress, *Int. J. Rock Mech. Min. Sci.*, 7, 557, 1970.

Byerlec, J. D., Friction of rocks, *Pure and Appl. Geophys.*, 116, 615–626, 1978.

Cao, T.-Q. and K. Aki, Seismicity simulation with a rate— and state—dependent friction law, *Pure Appl. Geophys.*, 124, 487–513, 1986.

Carlson, J. M., Time intervals between characteristic earthquakes and correlations with smaller events: An analysis based on a mechanical model of a fault, *Jour. Geophys. Res.*, 96, 4255–4267, 1991.

Carlson, J. M., and J. S. Langer, Properties of earthquakes generated by fault dynamics, *Phys. Rev. Lett.*, 67, 2632–2635, 1989.

Gabrielov, A. M., V. I. Keilis—Brook, T. A. Levshiva, and V. A. Shaposhnikov, Block model for dynamics of the lithosphere, *Computational Seismology*, 19, 168–178, 1986.

Langer, J. S. and C. Tang, Rupture propagation in a model of an earthquake fault, *Phys. Rev. Lett.*, 67, 1043–1046, 1991.

Ma, Z., G. Zhang, Z. Fu, *Nine Great Earthquakes in China*, Seis. Publ., Beijing, 1990.

Mogi, K., Magnitude frequency relation for elastic shocks accompanying fracture of various materials and some related problems in earthquakes, *Bull. Earthquake Res. Inst.*, 40, 831–853, 1962.

Nishenko, S. P., and R. Buland, A generic recurrence interval distribution for earthquake forecasting, *Bull. Seismol. Soc. Am.*, 77, 1382–1399, 1987.

Nussbaum, J. and A. Ruina, A two degree—of—freedom earthquake model with static / dynamic friction, *Pure and Applied Geophys.*, 125, 629–656, 1987.

Scholz, C., The frequency—magnitude relation of microfracturing in rock and its relation to earthquakes, *Bull. Seis. Soc. Am.*, 58, 399–415, 1968.

Shi, Y. and B. A. Bolt, Standard error of the magnitude—frequency b value, *Bull. Seis. Soc. Am.*, 72, 1677–1687, 1982.

Shimazaki, K. and T. Nakata, Time—predictable recurrence model for large earthquakes, *Geophys. Res. Lett.*, 7, 279–282, 1980.

Zhu, Y., and Y. Shi, Nonlinear dynamic modeling in seismicity analysis, *Chinese Journal of Geophysics*, 34, 27–40, 1991.

Yaolin Shi, Dept. of Earth Sciences, Graduate School, Academia Sinica, P.O.Box 3908, Beijing 100039, China.

Lumin Geng and Guomin Zhang, Center for Analysis and Prediction, State Seismological Bureau, P.O.Box 166, Fuxing Ave. 63, Beijing 100036, China.

Methods for Improving the Prediction of Dynamical Processes with Special Reference to the Atmospheric Circulation

JOHAN GRASMAN

Department of Mathematics, Agricultural University Wageningen, The Netherlands

PETER HOUTEKAMER

Division de Recherche en Prevision Numerique, Atmospheric Environment Service, Dorval, P.Q., Canada

Sensitive dependence on the initial state limits the predictability of future states of a nonlinear dynamical system. Methods for assessing and reducing the error in a prediction are discussed. In particular a variational data assimilation technique, an error growth analysis and a model validation procedure, known as the sentinel method, are reviewed. All three methods make use of the (adjoint) local tangent linear equation.

INTRODUCTION

This contribution deals with sensitive dependence on the initial state of nonlinear systems and the way it affects the predictability of the dynamics of such a system. A small change in the initial state may cause a large change in the course of a trajectory for the time that follows. It gives rise to chaotic dynamics of the system [*Bergé et al.*, 1984].

As mentioned sensitive dependence on the initial state complicates the prediction of future states of a system due to the limited accuracy by which one can estimate this initial state. The idea that a successful prediction of the far future can be made if sufficient information and computing capacity are available has turned out to be incorrect. Small scale perturbations may grow in a finite time to a full scale level and give rise to a horizon in our view of the future states of a system. In our study of predictability we take this restriction for given. An estimate for the growth of the error follows from the largest local Lyapunov exponent of the local tangent linear system that is obtained from linearization along the trajectory through the (estimated) initial state, see [*Yoden and Nomura*, 1993].

The purpose of this contribution is to discuss mathematical methods that help to reduce the error as far as possible and by which an estimate of the size of this error can be made. In the first place one should put all efforts in measuring as accurate as possible. Assuming that this is the case, we can use mathematical techniques to make an improved estimate of the initial state by incorporating observations and knowledge of previous states using the model of the dynamical system. This is known as a data assimilation method. Next we can make our prediction of the evolution in time of the system and include an analysis of the growth of the error. The result will be in the form of an approximation of the distribution of the error size.

At the end of this procedure it may turn out that we are not satisfied with the model we use to describe the physical phenomenon. However, it is difficult to decide whether the difference between prediction and eventual realization is due to shortcomings of the model or to sensitive dependence which cannot be avoided. This question can be brought closer to an answer by introduction of a functional known as the sentinel function being proposed by [*Lions*, 1988]. This function is an average with a weight chosen such that the dependence on the initial state is minimized.

Thus, with the view at the problem of making accurate predictions we present three topics: data assimilation, error growth estimation and model validation. These topics look rather diverse. However, there is a concept that binds them; that is the use of the adjoint local tangent linear system. This approach was first suggested by [*Marchuk*, 1974]. A series of articles by French geophysicists [*Le Dimet and Talagrand*, 1986; *Courtier and Talagrand*, 1987; and *Talagrand and Courtier*, 1987] initiated a wider use of this method especially in the atmospheric sciences [*Houtekamer*, 1992]. This linear approach implies that the application of the proposed method is restricted to short time periods and to synoptic scales. Diabatic processes such as precipitation are highly nonlinear. Consequently the applications of adjoint methods have sofar been restricted to the (large) synoptic scales where one may think that these processes may not be very important. The short-range weather is influenced by events on the synoptic scale. An example is the timing of the arrival of a low pressure system. This is determined by the phase of the synoptic scale wave. Such phase-errors can be described with linear methods.

In the section on the variational assimilation we summarize the results of this study, while in the next section we present a study [*Houtekamer*, 1991] on error growth based on the (adjoint) local tangent equation. In the final section we discuss the sentinel method of [*Lions*, 1990], where again the adjoint tangent linear system plays an important role.

Starting point of our analysis is a system of ordinary differential equations for the state vector $x(t) = (x_1(t), ..., x_n(t))$:

$$\frac{dx}{dt} = F(x) \quad \text{with} \quad x(t_0) = x_0 . \quad (1ab)$$

For the atmospheric circulation problem of [*Houtekamer*, 1991] this system acts as (finite dimensional) spectral approximation of the stream function of a baroclinic flow (two layers model). We assume that over the time interval (t_0, t_1) observation values of the state variables are given, while a forecast for $t_2 (> t_1)$ has to be made.

VARIATIONAL ASSIMILATION METHOD

The starting value x_0 of (1ab) will be varied in order to find a trajectory that comes as close as possible to the observation values. The equation of first variation,

$$\frac{d}{dt}\delta x = F'[t]\delta x, \quad \delta x(0) = \delta x_0$$

with

$$F'[t] = \left[\frac{\delta F_i}{\delta x_j}(x(t))\right]_{n \times n},$$

has a solution of the form

$$\delta x(t) = R(t,t_0)\delta x_0 . \quad (2)$$

For the trajectory $x(t)$ satisfying (1ab) we define a cost function for the total difference between $x(t)$ and the observations:

$$J(x(t)) = \int_{t_0}^{t_1} H(x(t),t)dt , \quad (3)$$

where $H(x(t),t)$ is a scalar measuring the distance between x and the observations available at time t (for instance, the squared norm of the difference between the observations at time t and the corresponding components of x).

Variation of the starting value changes the cost function by

$$\delta J = \int_{t_0}^{t_1} (\nabla_x H[t], \delta x(t))dt , \quad (4)$$

where (,) denotes the inner product induced by the kinetic energy (x,x). Using (2) we obtain

$$\delta J = \int_{t_0}^{t_1} (\nabla_x H[t], R(t,t_0)\delta x_0)dt$$

$$= \int_{t_0}^{t_1} (R^*(t,t_0)\nabla_x H[t], \delta x_0)dt,$$

so that

$$\nabla_{\delta x_0} J = \int_{t_0}^{t_1} R^*(t,t_0)\nabla_x H[t]dt . \quad (5)$$

The function $R^*(t, t_0)$ is as follows related to the adjoint local tangent linear equation

$$-\frac{d}{dt}\delta' x = F'^*[t]\delta' x, \quad \delta' x(t_1) = \delta' x_1 . \quad (6ab)$$

The solution of this equation reads

$$\delta' x(t) = P(t,t_1)\delta' x_1 \quad \text{with} \quad P(t,t_1) = R^*(t_1,t).$$

Moreover, the solution of

$$-\frac{d}{dt}\delta' x = F'^*[t]\delta' x + \nabla_x H[t], \quad \delta' x(t_1) = 0 \quad (7ab)$$

is of the form

$$\delta' x(t) = \int_{t}^{t_1} P(t,\tau)\nabla_x H[\tau]d\tau. \quad (8)$$

Combining (5) and (8) we obtain

$$\nabla_{\delta x_0} J = \delta' x(t_0).$$

Thus by one (backward) integration of (7) we find the gradient of the cost function as it depends on the starting value. The value of x_0 that minimizes $J(x(t))$, can then be found by e.g. the conjugated gradient method.

ESTIMATE OF ERROR IN FORECAST

Once we have determined the initial state $x(t_0) = x_0$ that minimizes the difference between the trajectory $x(t)$ and the observations over the interval $[t_0, t_1]$, we can continue the integration of the system to a future time t_2 ($> t_1$). We assume that at t_1 there is an error of average size ε. The deformation of a spherical initial distribution is investigated. Taking N starting values in state space with a n-dimensional normal distribution with expected value $x(t_1)$ and variance ε^2, we find after integration of the full nonlinear system up to $t = t_2$ N realizations $X_\varepsilon(t_2)$. From the statistics of $|X_\varepsilon(t_2) - x(t_2)|$ we can conclude about the distribution of the error in the forecast.

For the development of errorgrowth analysis in meteorology we refer to [*Lorenz*, 1965] and [*Farell*, 1989, 1990]. The papers by Farrell draw attention to the fact that finite-time errorgrowth analysis (optimal modes) is not very much related with asymptotic errorgrowth (Lyapunov exponents).

We now focus our attention on the use of local tangent linear equations for computing the error distribution in forecasts. Because of the linearization the result will only hold for weather forecasts having a maximal range of 3 days. For the linear system we obtain from (2)

$$\delta X(t_2) = R(t_2, t_1) \delta X_1 . \qquad (9)$$

For the error energy at $t = t_2$ we find

$$\rho^2 = (\delta X(t_2), \delta X(t_2)) = (R\delta X_1, R\delta X_1) = (R^*R\delta X_1, \delta X_1)$$

with $R = R(t_1, t_2)$. It is noted that $Q = R^*R$ is a symmetric matrix with positive eigenvalues $\mu_i = \lambda_i^2$, $i = 1, ..., n$. In [*Lorenz*, 1985] it is shown that for the expected value of the error energy at t_2 holds

$$E(\rho^2) = \frac{\varepsilon^2}{n} \sum_{j=1}^{n} \lambda_j^2 .$$

For error growth only the largest eigenvalues are important. These can be computed with the Lanczos algorithm. This is as follows. We take a random direction Δ_1 ($|\Delta_1| = 1$) and construct

$$\Delta_2' = Q\Delta_1 - (Q\Delta_1, \Delta_1)\Delta_1, \quad \Delta_2 = \Delta_2'/|\Delta_2'|.$$

We continue this Gram-Schmidt type of procedure up to Δ_j with $j \leq n$. Next the matrix

$$R_j = [R\Delta_1 \; R\Delta_2 \; ... \; R\Delta_j]$$

is introduced. It can be shown that the ½j largest eigenvalues of $R_j^* R_j$ are close to the first ½j eigenvalues of $Q = R^*R$. Let $v^{(1)}, ..., v^{(j)}$ be the corresponding eigenvectors. These are orthogonal because of the fact that Q is symmetric.

We now approximate the growth of the error from t_1 to t_2 by taking only the contribution coming from the eigenspace belonging to the first j eigenvectors $v^{(1)}, v^{(2)}, ..., v^{(j)}$. Initially a random realization has coordinates:

$$\delta X_\varepsilon(t_1) = \sum_{i=1}^{j} d_i v^{(i)} \varepsilon,$$

with d_i, $i = 1, ..., j$ random independent variables with expected value 0 en variance n^{-1}. An ensemble of vectors d represents an initially spherical distribution of errors. This ensemble becomes stretched at time t_2,

$$\delta X_\varepsilon(t_2) = \sum_{i=1}^{j} d_i R v^{(i)} \varepsilon .$$

For the error energy ρ^2 at time t_2 we obtain:

$$\rho^2 = \left(\sum_{i=1}^{j} d_i R v^{(i)}, \sum_{i=1}^{j} d_i R v^{(i)}\right)\varepsilon^2 = \left(\sum_{i=1}^{j} d_i R^* R v^{(i)}, \sum d_i v^{(i)}\right)\varepsilon^2 \qquad (10)$$
$$= \sum_{i=1}^{j} d_i^2 \lambda_i^2 (v^{(i)}, v^{(i)}) \varepsilon^2 = \sum_{i=1}^{j} d_i^2 \lambda_i^2 \varepsilon^2 .$$

From the ensemble of random vectors d and Eq. (10) we can estimate the probability distribution for the error ρ. This is done for a 100 day experiment. In this the forecast lengt $t_1 - t_2$ is kept constant and the time t_1 increases with steps of 1 day. Because $R(t_1, t_2)$ depends on the trajectory x between time t_1 and t_2 the eigenvalues are time dependent as well. In figure 1 we give, as a function of time, the probability P ($\rho < 2\varepsilon$) of a high quality forecast, the probability P ($\rho < 3\varepsilon$) of a forecast of medium or better quality and P ($\rho < 4\varepsilon$) which is the probability that a forecast is not of low quality.

In [*Barkmeijer and Opsteegh*, 1992] validations of local skill forecasts for the 2-day, 3-day and 4-day ECMWF prediction are given. Their method shows good performance for 2 and 3 days. [*Veyre*, 1992] has independently done comparable work with similar results. The linear method is found acceptable for 72 hours for the synoptic scales.

REMOVAL OF SENSITIVITY TO THE INITIAL STATE IN A MODEL VALIDATION

The initial value problem (1ab) represents a model of a physical system. We assume that, in case it is finite dimensional

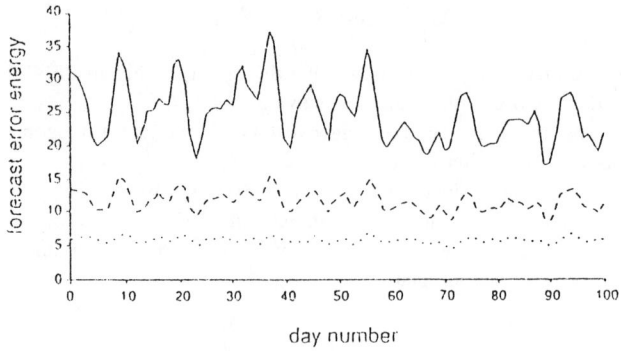

Fig. 1. Variations in time of the probabilities $P(\rho < 2\varepsilon)$, $P(\rho < 3\varepsilon)$ and $P(\rho < 4\varepsilon)$ during a 100 day integration. $P(\rho < 2\varepsilon)$ is indicated with circles; $P(\rho < 3\varepsilon)$ with squares; $P(\rho < 4\varepsilon)$ with triangles.

approximation of an infinite dimensional problem, the truncation n is taken sufficiently high. Then it still can be that in the model not all relevant processes are incorporated. Including these processes we may have in reality for $x(t; \lambda, \tau)$ a perturbed system of the form

$$\frac{dx}{dt} = F(x) + \lambda G(t), \quad x(t_0) = x_0 + \tau \xi_0, \quad t_0 < t \leq t_1 \quad (11)$$

with

$$\int_{t_0}^{t_1} |G(t)| dt = 1 \text{ and } |\xi_0| = 1.$$

Here λ represents the size of the (unknown) perturbation of the model equation and τ the inaccuracy in the initial state. We now adapt Lions' method of constructing a sentinel function for partial differential equations to systems of ordinary differential equations. It is remarked that the computations follow closely that of the variational assimilation method. The vector function $x(t; 0,0)$ satisfying the model equations (1ab) should agree with the observations. In order to make a comparison we introduce the average

$$A(\lambda,\tau) = \int_{t_0}^{t_1} (h(t), x(t;\lambda,\tau)) dt, \quad (12)$$

where the vector weight function $h(t)$ with

$$\int_{t_0}^{t_1} |h(t)| dt = 1, \quad h_i(t) \geq 0, \quad i = 1,\ldots,n,$$

follows from the observation procedure. It is noted that for λ and τ small the difference between the observation average and the model average is approximated by

$$A(\lambda, \tau) - A(0,0) \approx A_\lambda(0,0)\lambda + A_\tau(0,0)\tau.$$

Due to sensitive dependence on the initial state the coefficient $A_\tau(0,0)$ may be large, so that the second term dominates the difference. A reduction of this contribution can be achieved by taking the generalized average or sentinel function

$$S(\lambda,\tau) = \int_{t_0}^{t_1} (h(t) + w(t), x(t;\lambda,\tau)) dt \quad (13)$$

with $w(t)$ an appropriate weight vector such that

$$S_\tau(0,0) = \int_{t_0}^{t_1} (h(t) + w(t), x_\tau(t;0,0)) dt = 0. \quad (14)$$

Moreover, $w(t)$ should be as small as possible so that the weight function $h(t)$ of the observation process is the least affected. Thus $w(t)$ must be such that

$$I(w) = \frac{1}{2} \int_{t_0}^{t_1} (w(t), w(t)) dt \text{ is minimal}. \quad (15)$$

From (1ab) we derive the initial value problem for $x_\tau(t; 0,0)$

$$\frac{d}{dt} x_\tau = F'[t] x_\tau, \quad x_\tau(0) = \xi_0. \quad (16ab)$$

It is remarked that the initial error ξ_0 is unknown. This problem is overcome by switching to the adjoint local tangent linear equation

$$-\frac{dq}{dt} = F'^*[t]q + h(t) + w(t), \quad q(t_1) = 0 \quad (17ab)$$

with solution (see (7) - (8))

$$q(t) = \int_{t}^{t_1} P(t,\tau)\{h(\tau) + w(\tau)\} d\tau. \quad (18)$$

Similar as in (5) and (8) we have

$$S_\tau(0,0) = \int_{t_0}^{t_1} (h(t)+w(t), R(t,t_0)\xi_0) dt =$$
$$= \int_{t_0}^{t_1} (R^*(t,t_0)(h(t)+w(t)), \xi_0) dt = (q(t_0), \xi_0). \quad (14)$$

Thus, we must choose $w(t)$ such that $q(t_0) = 0$. The control problem for $w(t)$ satisfying (15) with constraint $q(t_0) = 0$ given by (18) is solved as follows. We introduce the Lagrange multipliers $\mu_1, ..., \mu_n$ and minimize

$$J(w) = \frac{1}{2}\int_{t_0}^{t_1}(w(t),w(t))dt + (\mu,q(t_0)).$$

Using (18) we have

$$J(w) = \int_{t_0}^{t_1}\frac{1}{2}(w(t),w(t)) + (\mu,P(t_0,t)\{h_0(t)+w(t)\})dt =$$
$$\int_{t_0}^{t_1}\frac{1}{2}(w(t),w(t)) + (P^*(t_0,t)\mu,\{h(t)+w(t)\})dt,$$

so that $\delta J/\delta w = 0$ for

$$w(t) = -P^*(t_0,t)\mu = -R(t,t_0)\mu.$$

From (18) we then have the condition

$$q(t_0) = \int_{t_0}^{t_1}P(t_0,t)\{h(t)-R(t,t_0)\mu\}dt = 0.$$

Consequently, the Lagrange multipliers satisfy

$$\mu = \left\{\int_{t_0}^{t_1}P(t_0,t)R(t,t_0)dt\right\}^{-1}\left\{\int_{t_0}^{t_1}P(t_0,t)h(t)dt\right\}.$$

Assuming that $S_\lambda(0,0) \neq 0$, the difference between $S(\lambda, \mu)$ and $S(0,0)$ yields an order estimate of the error in the solution from the perturbation $\lambda G(t)$ in the differential equation:

$$S(\lambda,\mu) - S(0,0) = \lambda S_\lambda(0,0) =$$
$$\int_{t_0}^{t_1}(h(t) + w(t),\lambda x_\lambda(t;0,0))\,dt.$$

If $|w(t)|$ is not of a larger magnitude than $h(t)$, then the error in the solution is estimated by

$$\lambda x_\lambda(t; 0,0) = O(|S(\lambda,\mu) - S(0,0)|).$$

In [*Mous and Grasman*, 1993] the method is applied to a low order spectral model of the atmospheric circulation in a beta-plane. There the results are compared with those of a more traditional method, the extended Kalman filter.

References

Barkmeijer, J., and J.D. Opsteegh, Local skill prediction with a simple model, in *Proceedings of a workshop held at ECMWF on New developments in Predictability*, November 1991, Reading, pp. 55-64, 1992.

Bergé, P., Y. Pomeau and C. Vidal, Order within Chaos: Towards a deterministic approach to turbulence, Wiley, New York, 1984.

Courtier, P., and O. Talagrand, Variational assimilation of meteorological observations with the adjoint vorticity equation II: Numerical results. *Quarterly Journal Royal Meteorological Society, 113*, pp. 1329-1347, 1987.

Farrell, B.F., Optimal excitation of baroclinic waves, *J. Atmos. Sci. 46*, pp. 1193-1206, 1989.

Farrell, B.F., Small error dynamics and the predictability of atmospheric flows, *J. Atmos. Sci. 47*, pp. 2409-2416, 1990.

Houtekamer, P., Variation of the predictability in a low order spectral model of the atmospheric circulation, *Tellus 43A*, pp.177-190, 1991.

Houtekamer, P., Predictability in models of the atmospheric circulation, Thesis, Agricultural University Wageningen, The Netherlands, pp. 121, 1992.

Le Dimet, F.X., and O. Talagrand, Variational algorithms for analysis and assimilation of meteorological observations: theoretical aspects. *Tellus 38A*, pp. 97-110, 1989.

Lions, J.L., Sur les sentinelles des systmes distribus, *C.R.A.S., Paris 307*, pp. 819-823, pp. 865-870, 1988.

Lions, J.L., Sentinels and stealhy perturbations, *Proc. Int. Symp. on Assimilation of Observations in Meteorology and Oceanography*, Clermond-Ferrand, World Meteorological Organization, pp. 13-18, 1990.

Lorenz, E.N., A study of the predictability of a 28-variable atmospheric model, *Tellus 17*, pp. 321-333, 1965.

Lorenz, E.N., The growth of errors in prediction, in *Turbulence an Predictability in Geophysical Fluid Dynamics and Climate Dynamics*, North Holland, Amsterdam, pp. 243-265, 1985.

Marchuk, G.I., Numerical solution of the problems of the dynamics of the atmosphere and the ocean, *Gidrometeo*, Leningrad, 1974.

Mous, S.L., and J. Grasman, Two methods for assessing the size of external perturbations in chaotic processes, *Mathematical Models and Methods in Applied Sciences, 3*, pp. 577-593, 1993.

Talagrand, O., and P. Courtier, Variational assimilation of meteorological observations with the adjoint vorticity equation. Part 1: Theory, *Quart. J. Roy. Meteo. Soc. 113*, pp.1311-1328, 1987.

Veyre, P., Direct prediction of error variances by the tangent linear model: a way to forecast uncertainty in the short range?, in *Proceedings of a workshop held at ECMWF on New developments in Predictability*, November 1991, Reading, pp. 65-86, 1992.

Yoden, S. and M. Nomura, Finite-time Lyapunov stability analysis and its application to atmospheric predictability, preprint, Dept. of Geophysics, Kyoto University, to appear in J. Atmos. Sci., 1993.

J. Grasman, Dept. of Mathematics, Agricultural University, Dreijenlaan 4, 6703 HA Wageningen, The Netherlands.

P.L. Houtekamer, Division de Recherche en Prevision Numerique, Atmospheric Environment Service, 2121 Transcanadian Highway, Dorval, P.Q. H9P1J3, Canada.

The Nonlinear Asymptotic Stage of the Rayleigh-Taylor Instability with Wide Bubbles and Narrowing Spikes

V. M. CHERNIAVSKI AND YU. M. SHTEMLER

Institute of New Technologies, Moscow, Russia

The potential flow of an incompressible inviscid heavy fluid over a light one is considered. The integral version of the method of matched asymptotic expansion is applied to the construction of the solution over long intervals of time. The asymptotic solution describes the flow in which a bubble rises with constant speed and the "tongue" is in free fall. The outer expansion is stationary, but the inner one depends on time.

It is shown that the solution exists within the same range of Froude number obtained previously by Vanden-Broeck (1984a,b). The Froude number and the solution depend on the initial energy of the disturbance. At the top of the bubble, the derivative of the free-surface curvature has a discontinuity when the Froude number is not equal to 0.23. This makes it possible to identify the choice of the solution obtained in a number of studies with the presence of an artificial numerical surface tension. The first correction term in the neighborhood of the tongue is obtained when large surface tension is included.

INTRODUCTION

For two-layered fluid media over long intervals of time, rising bubbles and falling tongues have been observed in a number of laboratory experiments [Emmons, 1960]. It is characterized by a constant dimensionless speed of a wide bubble (Froude number $Fr = 0.2 - 0.3$) and a constant acceleration of a narrowing tongue.

As usual, the stationary solution is represented as a sum of a singular and a regular series [Birkhoff and Carter, 1957]. Birkhoff and Carter (1957) and Vanden-Broeck (1984a,b) have lost the second term in the singular series. To obtain the correct (nonoscillatory) solution and to preserve the energy conservation law, it is necessary to keep this second term.

Instead of the inner expansion, we use the conditions at the jump between the branches of the steady solution. These conditions are obtained using the conservation laws, which give the correct asymptotic formulas for a steady solution. In addition, the outcome of these are employed to estimate numerical errors. We show that the solution exists within the same range of Froude number (Fr) as obtained previously [Vanden-Broeck, 1984a]. We find that Fr depends on the initial energy of the disturbance. At the top of the bubble, the derivative of the free-surface curvature has a discontinuity when $Fr \neq 0.23$. This makes it possible to identify the choice of solution obtained in a number of studies with the presence of an artificial numerical surface tension. We obtain the first correction term in the neighborhood of the tongue when large surface tension is included.

FORMULATION OF THE PROBLEM

The potential flow of an inviscid incompressible heavy fluid lying above a light one is investigated. In a Cartesian coordinate system XY, let the plane $Y = 0$ be the unperturbed free surface of a heavy liquid ($Y \geq 0$). Let us consider a two-dimensional potential flow of the fluid with unit density. We assume at long times T that the flow is spatially periodic in X and symmetric with respect to the Y axis

$$\begin{aligned}
\Delta \Phi &= 0, \\
\Phi(X + \lambda, Y, T) &= \Phi(X, Y, T), \\
\Phi(-X, Y, T) &= \Phi(X, Y, T), \\
\frac{d\Gamma}{dT} &= 0 \quad (\Gamma = 0), \\
\Gamma(X + \lambda, Y, T) &= \Gamma(X, Y, T), \\
\Gamma(-X, Y, T) &= \Gamma(X, Y, T), \\
P &= 0 \quad (\Gamma = 0),
\end{aligned} \qquad (1)$$

RAYLEIGH-TAYLOR INSTABILITY

$$P = -\left(Y + \frac{1}{2}\left|\frac{dF}{dZ}\right|^2 + \frac{\partial \Phi}{\partial T}\right),$$
$$Z = X + iY, \quad F = \Phi + i\Psi,$$
$$\left|\frac{dF}{dZ}\right| \to 0, \quad (Y \to \infty) \quad .$$

Here P is the pressure, F is the complex potential, Δ is the Laplacian operator, and $\Gamma(X, Y, T) = 0$ is the equation of the free surface. The pressure outside the liquid is assumed equal to zero. To transform to dimensionless variables, we shall take the wavelength λ and the free fall acceleration g as characteristic scales. The flow is shown schematically in Figure 1. (The dashed curves show the time evolution of the free surface.) We assume that the bubbles rise when $X = \pm 1/2$, while a tongue of heavy liquid collapses when $X = 0$. Assuming that the coordinate system moves with the velocity of the bubble top, $w = Fr$, we have

$$T = t - t_0, \quad Z = iwt + z + iy_0,$$
$$F(Z, T) = -iwz + f(z, t) + \Phi_0(t),$$
$$P(X, Y, T) = p(x, y, t), \quad (2)$$
$$\Gamma(X, Y, T) = \gamma(x, y, t),$$
$$z = x + iy, \quad f = \phi + i\psi \quad .$$

Here, $\Phi_0(t) = -w(t^2 - wt)/2 - y_0 t + \phi_0$. We assume that the initially unknown constant w_0 is the velocity of the bubble top, when t tends to infinity. We define the dimensionless value w in terms of the dimensional velocity w_0 as

$$w = \frac{w_0}{\sqrt{\lambda g}} \quad .$$

We now transform one period of the flow on the z-plane onto a halfstrip on the q-plane

$$q = s + ir, \quad r > 0, \quad |s| \leq \pi,$$
$$\frac{\partial x}{\partial s} = \frac{\partial y}{\partial r}, \quad \frac{\partial x}{\partial r} = -\frac{\partial y}{\partial s},$$
$$x(-s, r, t) = -x(s, r, t), \quad x(\pi, r, t) = \frac{1}{2}, \quad (3)$$
$$\gamma(x(s, 0, t), y(s, 0, y), t) = 0 \quad |s| \leq \pi,$$
$$\frac{\partial y}{\partial r} \to \frac{\pi}{2}, \quad r \to \infty \quad .$$

The Form of the Steady-State Solution in the Absence of Surface Tension

Let us consider the steady-state solution f, z of the system of equations (1)–(3), which is an outer solution in the asymptotic expansion in $1/t$ which is valid outside the region of the narrow tongue. According to Birkhoff and Carter [1957] we have

$$f = \frac{w}{2\pi} \ln \frac{\theta}{(1-\theta)^2} - wy_\infty \quad (4)$$
$$\theta = \exp(iq), \quad y_\infty = \text{const.}$$

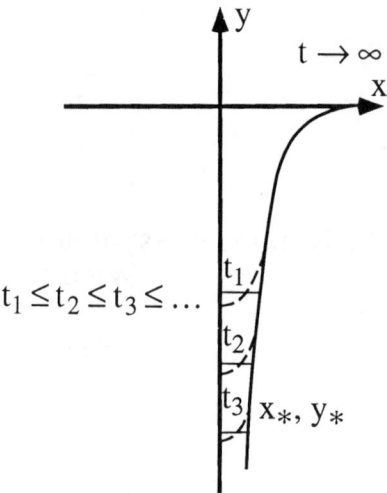

Figure 1. The flow is shown schematically. The dashed curves show the time evolution of the free surface. The bubbles rise at $X = \pm 0.5$, the tongue of heavy liquid collapses at $X = 0$.

The free surface is described by a curve which everywhere is analytic, except at points where the velocity is infinite ($\theta = 1$) and, possibly, at points with zero velocity ($\theta = -1$) [Birkhoff, 1957].

The unknown z is sought in the form

$$iz = L^{2/3} \sum_{k=0}^{n} D_k L^{-k} + \sum_{k=0}^{N} iz_k \theta_k + \frac{i}{2\pi} q$$
$$\left(L = \ln\left(\frac{1-\theta}{2}\right)^2 - 1\right) \quad (5)$$

This solution includes D_1 and differs from Birkhoff (1957) and Vanden-Broeck (1984a). The coefficients D_k for $k = 1, 2, \cdots$ are determined by using an asymptotic expansion procedure in powers of L as $\theta \to 1$, and $D_0 = (9w^2/32\pi^2)^{1/3}$, while z_k is found from the dynamic condition on the free surface with the aid of the Fast Fourier Transform and Newton's method.

Conservation Laws in the Absence of Friction

Due to the symmetry of the problem and the conditions at infinity, the conservation laws for the system of equations (1) and (2) can be written in the form

$$\frac{di_1}{dt} = -\frac{w}{2}\phi_\infty, \quad i_1 = \frac{1}{2}\int_S (\phi d\psi - y^2 dx)$$
$$\frac{di_2}{dt} = w, \quad i_2 = -\int_S y dx,$$
$$\frac{di_3}{dt} = -i_2 - \frac{w^2}{2}, \quad i_3 = -\int_S \phi dx,$$
$$\frac{di_4}{dt} = i_3 + \phi_\infty, \quad i_4 = -\frac{1}{2}\int_S y^2 dx, \quad (6)$$

$$\frac{di_5}{dt} = 4i_1 - 7i_4 + b(t), \quad i_5 = \int_S \phi(ydx - xdy),$$

$$b = \frac{1}{2}\int_h^\infty (\phi_y^2 - w^2)\big|_{s=\pi} dy + \phi_\pi h_t - \frac{1}{2}h^2 - \frac{1}{2}w^2 h \quad .$$

Here, $h(t)$ is the ordinate of the bubble top with respect to the moving coordinate system, $\phi_\infty(t)$ is an arbitrary function depending on time. The integrals $i_1, i_2, \cdots i_5$ represent the deviations of the total energy, mass, momentum, potential energy, and "virial" from the corresponding values in the unperturbed state $\phi_\pi = \phi(\pi, 0, t)$.

Let us examine the behavior of the nonstationary solution as it approaches the stationary solution, equations (4) and (5), outside the narrow tongue region. It can be shown, up to exponentially small terms and parameters y_0 and ϕ_0, that

$$\phi_\infty(t) \to 0, \quad h(t) \to 0, \quad dh/dt \to 0,$$
$$b(t) \to \int_0^\infty (\phi_y^2 - w^2)\big|_{s=\pi} dy \quad (t \to \infty) \quad . \quad (7)$$

The order of the nonstationary solution in the neighborhood of the tongue is obtained from Eq. (6) using Eq. (7), namely

$$x \sim t^{-1}, \quad y \sim t^2, \quad \phi \sim t^3, \quad \psi \sim 1 \quad (t \to \infty) \quad . \quad (8)$$

Nonstationary Jump Conditions in the Neighborhood of the Tongue

The solution in the neighborhood of the tongue can be described by an integral along a contour which approximates the free surface. (In Figure 1, the smooth curve corresponds to the stationary solution, while the dot-dashed curves represent the nonstationary jump; $x_*(t)$ and $y_*(t)$ are the coordinates of the right side of the jump at the time t). The even functions are discontinuous and the odd ones are continuous:

$$y(s, 0) = y_*(x_*), \quad x(s, 0) = \pm x_*,$$
$$p(s, 0) = p_*(x_*), \quad \phi(s, 0) = \phi_*(x_*), \quad (9)$$
$$\psi(s, 0) = \pm\psi_*, \quad s = \pm s_* \quad .$$

For $r = 0$ and $s \to 0$ $(t \to \infty)$ from Eqs. (4), (5) and (8) we obtain

$$y_*(x_*) = -\frac{w^2}{8x_*^2} + O(x_*^4),$$
$$p_*(x_*) = O(x_*^4),$$
$$\phi_*(x_*) = \left(\frac{w}{x_*}\right)^3 \frac{1}{24} + 2\frac{k}{w} + O(x_*^3), \quad (10)$$
$$\psi_*(x_*) = \frac{w}{2} + O(x_*^6),$$
$$x_*(t) = \frac{\mu}{t} \quad .$$

Consequences of the Conservation Laws for the Stationary Case

The proportionality constant μ is determined by the conservation laws. Using Eqs. (9) and (10), we evaluate the integrals i_k in Eq. (6) for the limiting solution of the nonstationary problem when $t \to \infty$. The solution is determined by solving Eq. (5) for the stationary problem and Eqs. (9) and (10), namely

$$i_1 = v + k, \quad i_2 = m + \frac{w^2}{2x_*},$$
$$i_3 = j - \frac{(w/2)^3}{x_*^2}, \quad i_4 = v - \frac{(w/2)^4}{3x_*^3}, \quad (11)$$
$$i_5 = d + \frac{7(w/2)^5}{12x_*^4} + \frac{2kw}{x_*} \quad .$$

Here, $v + k$, m, j, v and d are the regular parts of the integrals i_k with $k = 1, \cdots, 5$, and

$$m = -2\int_0^{1/2} [y - y_*(x)]dx - \frac{w^2}{2} + O(x_*^5),$$
$$v = -\int_0^{1/2} [y^2 - y_*^2(x)]dx - \frac{w^4}{24} + O(x_*^3),$$
$$j = -2\int_0^{1/2} [\phi - \phi_*(x)]dx + \frac{w^3}{6} - 2\frac{k}{w} + O(x_*^4),$$
$$k = -\frac{w^2}{2}\left(y_\infty + \frac{L_\infty - 3D_1/2}{2\pi}\right),$$
$$y_\infty = z_0 + \sum_{\ell=0} D_\ell L_\infty^{2/3-\ell},$$
$$L_\infty = 1 + \ln 4.$$

Integrating Eq. (6) with respect to time and taking into consideration Eq. (11), we obtain:

$$\mu = \frac{w}{2} \quad (12)$$

$$m = -\frac{w^2}{2}, \quad j = 0, \quad v = \frac{b_c}{3} \quad (13)$$

$$c_1 = v + k, \quad c_2 = -\frac{w^2}{2}, \quad c_3 = 0 \quad . \quad (14)$$

Here the c_k, with $k = 1, 2, 3$, are the initial values of the i_k at $t = 0$. According to Eqs. (6) and (7), the integrals i_4 and i_5 depend on the transition process in the first order approximation, and the equations for them have not been expressed in this approximation. According to Eq. (10), the value of μ in Eq. (12) determines the average acceleration of the tongue, which equals the acceleration of free fall. The integral identities (13) are consequences of the conservation laws for the stationary case. Equations (14) determine the particular initial data corresponding to the limiting solution.

Froude Number Dependence on the Initial Total Energy

From Eqs. (6) and (14), we obtain the dependence of the initial data $c_k^0 = i_k(t_0, c_n)$ on the parameter t_0:

$$c_1^0 + \frac{w}{2}\phi_\infty(t_0) = v + k,$$
$$c_2^0 = wt_0 - \frac{w^2}{2}, \quad c_3^0 = -\frac{wt_0^2}{2} \quad (15)$$

For comparison with physical experiments and numerical simulations, we can use Eq. (2) to obtain the relationship between the initial data with respect to the moving and fixed coordinate system ($t = t_0$):

$$E = c_1^0 - w\phi_\infty/2 + c_2^0(Y_0 + w^2/2) + wc_3^0 - Y_0^2/2,$$
$$M = c_2^0 - Y_0, \quad J = c_3^0 + wc_2^0 - \Phi_0. \quad (16)$$

Here $Y_0 = y_0 + wt$; E, M and J are the initial deviations in the total energy, mass and momentum with respect to the fixed coordinate system. These deviations are obtained by formal replacement of x, y, ϕ, and ψ by X, Y, Φ and Ψ for the i_k defined in Eq. (6). Without loss of generality we set $M = 0$ and $J = 0$ and, thereby choose a system of coordinates and the constant for the potential. On eliminating c_k^0 from Eqs. (15) and (16), we obtain the dependence of the parameters w, y_0 and ϕ_0 on the initial data, namely

$$E = v(w) + k(w) - \frac{w^3}{8},$$
$$y_0 = -\frac{w^2}{2}, \quad \phi_0 = -\frac{w^3}{2}. \quad (17)$$

Calculation Results

As a result of the calculations, it was found that a surface with a smooth curvature exists only for a certain value of $D_1 = D_*$. When $D_1 \neq D_*$, the solution oscillates and the identities (13) are no longer satisfied. A stationary solution corresponding to D_* has been found for Froude numbers in the range $0 < w < 0.37$. This interval is approximately the same as that has been found previously [Vanden-Broeck, 1984a]. Figure 2 shows the shape of the free surface $y = y(x)$, for three values of w. The derivative of the curvature of the interface surface at the top of a bubble has a discontinuity over the entire range of Froude numbers, except at one value ($w = 0.23$). Figure 3 shows the curvature $K = K(x)$ for three values of w.

We find that two possible values for the velocity of a rising bubble can exist for a single initial energy. Figure 4 shows the function $w(E)$ given by Eq. (17).

Small Amount of Surface Tension and Discontinuity of the Curvature Derivative

When the surface tension with coefficient σ is taken into account, the dynamic condition at the free surface $y = y(x)$ has the form

$$y(x) + \frac{1}{2}\left|\frac{f}{z}\right|^2 + \sigma K = \sigma K_0,$$

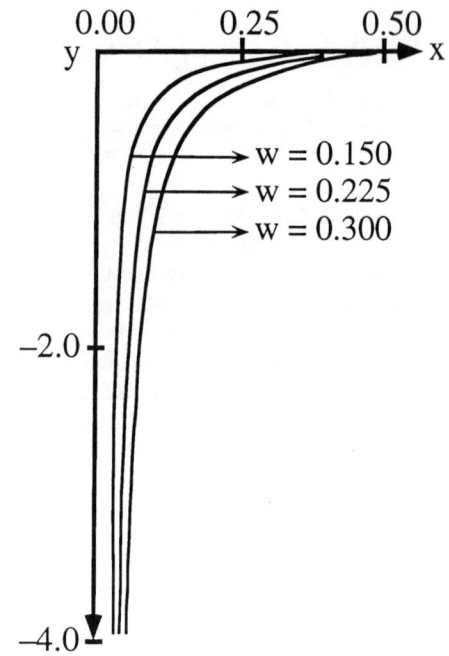

Figure 2. The shape of the free surface $Y = y(x)$ for three values of the Froude number, namely $w = 0.150$, 0.225, and 0.300.

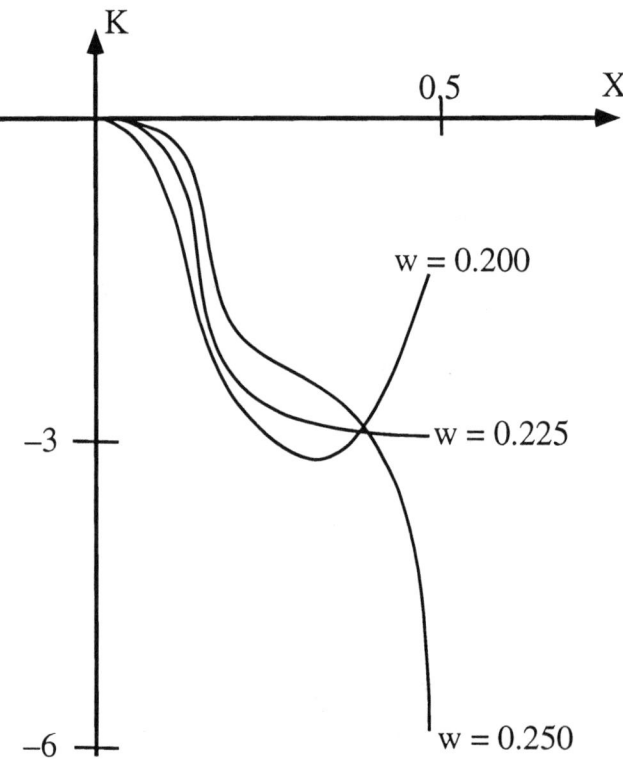

Figure 3. The curvature $K = K(x)$ for three values of the Froude number, namely $w = 0.200$, 0.225, 0.250.

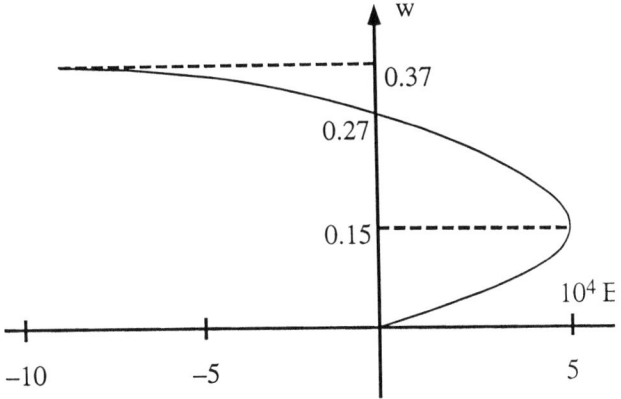

Figure 4. The dependence of Froude number upon the initial energy of the disturbances $w = w(E)$ given by Eq. (17).

$$f_z = \frac{df}{dz}, \quad K(x) = \frac{d}{dx}\left(\frac{y_x}{\sqrt{1+y_x^2}}\right), \quad (18)$$

$$K_0 = K\left(\frac{1}{2}\right), \quad y_x = \frac{dy}{dx}.$$

If $\sigma \neq 0$ the contact angle at the top of a bubble at the wall must be specified. Let

$$\frac{dy(x)}{dx} = 0 \qquad \left(x = \pm\frac{1}{2}\right). \quad (19)$$

We seek a solution of Eq. (18) in the form

$$y(x) = y^0(x) + \sigma y^1(x) + \cdots,$$
$$f_z = f_z^0 + \sigma f_z^1 + \cdots,$$
$$K = K^0 + \cdots . \quad (20)$$

It can be shown that the boundary condition (19) is satisfied in the zeroth order of σ and is violated in the next order, since $y_x^1 = K_x^0$ has a discontinuity at the top of the bubble for all w except $w = 0.23$.

A small amount of surface tension may, therefore, isolate a fully defined limiting velocity at which the bubble will rise and, thus, explain the results obtained by Baker et al. (1980).

The First Correction Term Due to a Large Surface Tension Coefficient

When the surface tension coefficient σ is not small, the first correction term in the steady-state solution of Eq. (18) is also proportional to σ. It can be seen from the conservation laws that if the surface tension is taken into account it is necessary to add the term $-\sigma \int_S \left(\sqrt{1+y_x^2} - 1\right) dx$ to the total energy i_1 and to exclude the law for the "virial"—in Eq. (6) [Benjamin, 1982].

Let us take only the first term in Eq. (5), namely

$$iz^0 = D_0 L^{2/3} .$$

It is easy to see that kinetic and potential energies in i_1 are of order t^3 and that the newly added term is of order t^2. Thus the term iz^0 is the main term in asymptotic expansion in the neighborhood of the tongue and the expansion (2) is also correct when σ is not small.

It can be shown that the linearized dynamic condition (18) has the form

$$\Re\left\{iz^1 + 2\frac{d(iz^1)}{d\ln(iz^0)}\right\} = \sigma K^0 - \sigma K_0^0 . \quad (21)$$

This means that the real part of a harmonic function, say $f(q)$, is equal on the surface to $\sigma K^0 - \sigma K_0^0$. Here K^0 is the curvature of the zeroth order surface. Let us map the half plane $q = s + ir, r \geq 0$ onto the circle $\omega = \rho \exp(i\alpha)$, $\rho \leq 1$, $q = i\frac{1-\omega}{1+\omega}$ and include the function $L_0(\omega) = \ln\frac{1+\omega}{1-\omega}$ taking the branch

$$L_0\left(e^{i\alpha}\right) = \ln\left|\cot\frac{\alpha}{2}\right| + i\frac{\pi}{2}\text{sign}(\alpha) .$$

It is evident that the function $L(q)$ in Eq. (5) is equal to $-2L_0(\omega)$ with asymptotic accuracy.

We construct a Schwarz function $f(q) = F(\omega)$ using the next theorem [Pyckteev, 1982]. If the density $f(s) = F(e^{i\alpha})$ is an even function of α and $|\alpha| \leq \pi$, then let us select the function which is equal to density for $0 < \alpha \leq \pi$, namely

$$\phi(\zeta) = \phi_1(\zeta) + \phi_2(\zeta),$$
$$\phi(\zeta) = F(e^{i\alpha}), \quad \zeta = e^{i\alpha},$$

where ϕ_1 is an even function, and ϕ_2 is an odd one, for $|\alpha| \leq \pi$. Then, the Schwarz function is

$$F(\omega) = \phi_1(\omega) - i\frac{2}{\pi}\phi_2(\omega)L_0(\omega) .$$

It is observed from Eq. (21) that

$$\phi_1(\zeta) = -\sigma K_0^0,$$

$$\phi_2(\zeta) = \frac{\sigma}{iz_q^0}\frac{1}{2i}\left\{\frac{d}{dq}\ln\left(iz_q^0\right) - \left[\frac{d}{dq}\ln\left(iz_q^0\right)\right]^*\right\}$$

and it is easy to show that, with asymptotic accuracy,

$$\phi_2(\omega) = \frac{\sigma}{4iD_0}2^{-2/3}L_0^{-5/3}(\omega)\left[L_0(\omega) - L_0\left(\frac{1}{w}\right)\right],$$

$$F(\omega) = -\frac{\sigma}{4\pi D_0}[2L_0(\omega)]^{-2/3}[L_0(\omega) - \bar{L}_0(\omega)] - \sigma K_0^0,$$

$$\bar{L}_0(\omega) = L_0\left(\frac{1}{\omega}\right),$$

$$f(q) = \frac{\sigma}{4\pi D_0}[2L]^{-2/3}(q)[L(q) - \bar{L}(q)] - \sigma K_0^0,$$

$$\bar{L}(q) = -2\bar{L}_0(\omega),$$

here the symbol * defines complex conjugation. Returning to Eq. (21), we have

$$\frac{1}{2}f(q) = \frac{d(iz^1)}{d\ln(iz^0)} + \frac{1}{2}iz^1$$
$$= \frac{\sigma}{2}\left[\frac{1}{4\pi D_0}L^{-2/3}(L - \bar{L}) - K_0^0\right] \quad . \tag{22}$$

From the ordinary differential equation (22) we obtain

$$iz^1(q) = \frac{1}{\sqrt{iz^0}}\frac{1}{2}\int \frac{iz_q^0}{\sqrt{iz^0}}f(q)dq$$

and on the free surface we have

$$iz^1(s) = -i\frac{\sigma}{2D_0}L^{-2/3}(s)\,\text{sign}(s) - \sigma K_0^0 \quad .$$

So, the first correction term is obtained for the solution of the problem which accounts the surface tension.

Discussion

An important issue is the suitability of this two-dimensional model of flow to the real three-dimensional one. The nonlinear problem requires special treatment for three-dimensional flow and does not reduce to the two-dimensional one, as in the linear case. We have investigated this in previous work [Gertsenshtein and Cherniavskii, 1985] and our results have shown that the solution does depend on the initial values and, in particular, on some aspects of three-dimensionality. But the two-dimensional case is justified because it survives under some conditions within the framework of the three-dimensional problem. On the other hand the results of the present work demonstrate that the two-dimensional solution discussed here can exist only if the initial values have the energy within some range. So, we have considered the solution to the three-dimensional problem which satisfies a special class of initial conditions and survives under some requirements.

A second issue which emerges is the relevance of these simulations to natural phenomena. Flows which are similar to our solutions have been observed in a set of laboratory experiments—see Emmons (1960), for example. Establishing the kind of initial conditions that can occur in nature is beyond our present considerations, but should be explored in the future. However, we believe that the determination of the precise solution to this nonlinear problem is a step toward understanding such natural phenomena as mantle density differentiation governed by gravity forces. This model could be regarded as a possible process in the mantle.

Conclusions

Asymptotic solutions which describe the behavior of a rising bubble of light fluid and of a narrow, falling tongue of heavy fluid over long times for development of the Rayleigh-Taylor instability have been obtained. It has been shown that the limiting upward velocity of a bubble (the Froude number) depends on the initial energy of the perturbations. The derivative of the curvature of the interface at the top of the bubble is discontinuous for all values except one, corresponding to a special value of the Froude number.

The first correction term in the neighborhood of the tongue has been obtained when large surface tension was taken into account. The regular part of stationary solution was obtained by numerical procedure using the Fast Fourier Transform and Newton's method.

References

Babenko K. I., and B. Yu. Petrovich, "Numerical investigation of the Rayleigh-Taylor instability problem," *Dokl. Akad. Nauk SSSR*, 255, no. 2, 318, 1980 (*Sov. Phys. Dokl.*, 25, 909, 1980).

Baker G. R., D. I. Meiron, and S. A. Orszag, "Vortex simulations of the Rayleigh-Taylor instability," *Phys. Fluids*, 23, 8, 1485, 1980.

Benjamin T. B., and P. J. Olver, "Hamiltonian structure, symmetries and conservation laws for water waves," *J. Fluid Mech.*, 1, 137, 1982.

Birkhoff G., and D. Carter, "Rising plane bubbles," *J. Math. Mech.*, 6, 769, 1957.

Davydov J. M., A. Yu. Dem'yanov, and G. A. Tsvetkov, "Numerical modelling of stabilization and consolidation of harmonics through the Rayleigh-Taylor instability by the particle-in-cell method" (in Russian), *Soobschenie po prikladnoi matematike VTS AN SSSR*, 1987.

Emmons H. W., C. T. Chang, and B. S. Watson, "Taylor instability of finite surface waves,"*J. Fluid Mech.*, 7, 177, 1960.

Garabedian P. R., "On steady-state bubbles generated by Taylor instability," *Proc. R. Soc. London, A241*, 423, 1957.

Gertsenshtein S. Ya., and Chernyavskii V. M., "On the nonlinear development of two-dimensional and three-dimensional perturbations in the presence of Rayleigh-Taylor instability," *Izv. Akad. Nauk SSSR, Mekh. Zhidk. i Gaza*, 2, 38, 1985.

Pyckteev G. N., "Priblizhennye metody vychisleniya integralov tipa Koshi specialnogo vida" (in Russian), 127pp., Nauka, Novosibirsk, 1982.

Vanden-Broeck J. M., "Bubbles rising in a tube and jets falling from a nozzle," *Phys. Fluids*, 27, 1090, 1984a.

Vanden-Broeck J. M., "Rising bubbles in a two-dimensional tube with surface tension," *Phys. Fluids*, 27, 2604, 1984b.

Volkova R. A., L. V. Kruglyakova, N. V. Michailova, et al., "Modeling of Rayleigh-Taylor instabilities in a three-dimensional incompressible fluid" (in Russian), Preprint, *Institut Pricladnoi Matematiki AN SSSR*, 86, 1985

V. M. Cherniavski and Yu. M. Shtemler, Institute of New Technologies, Kirovogradskaya 11, Moscow, 113587, Russia

Correlation Dimension of the Strange Attractor for Geomagnetic Field Variations

YU.S. TYUPKIN AND A.YA. FELDSTEIN

Geophysical Center, Russian Academy of Sciences, Moscow

We discuss the consequences of the hypothesis that chaotic behavior in the variations of the geomagnetic field (VGF) in the auroral zone can be described by a nonlinear dynamical system with a strange attractor. The correlation dimension of the strange attractor of the underlying dynamical system is regarded as a characteristic of chaotic behavior of the VGF and the time evolution of the correlation dimension is studied based on the VGF measured by auroral zone observatories. The seasonal and latitudinal variations of the correlation dimension are evaluated.

Introduction

As noted in recent years, some dynamical systems can admit bounded nonperiodic solutions whose behavior appears random even though no random quantities appear in these systems. Such systems are said to have chaotic dynamics or exhibit deterministic chaos. There are many examples of natural processes with chaotic dynamics (Lorenz, 1963; Ghil et al., 1985; Nicolis and Nicolis, 1984, 1987; Grassberger, 1984; Tsonis and Elsner, 1988; Moon, 1987; Mandelbrot, 1982; Anderson et al., 1988; Rapp et al., 1987).

The underlying objective was to reduce the description of the temporal evolution of the physical system, the dynamics of which is governed by partial differential equations and in principle requires an infinite number of variables, to an investigation of a few coupled nonlinear ordinary differential equations. To do this, the fields of the original physical system would be expanded according to a complete set of functions. The amplitudes of these functions then obey an infinite set of ordinary differential equations. It may happen, not far from equilibrium, that the system will be in a stable stationary state, i.e. small perturbations around that state eventually decay to zero. It may happen also, upon displacing the system further from equilibrium, that only a few of modes will become unstable and develop non-trivial nonlinear dynamics. All other modes are dynamically irrelevant and are dominated by few relevant modes. These reasons are used often to justify the preceding methodology for processes with regular evolution. Examples of systems with chaotic dynamics in low dimensions demonstrate that similar arguments can be successful for the investigation of natural processes with seemingly random time histories.

Considerable progress in the understanding of low-dimensional chaos was achieved by the quantitative characterization of the post-transient phase-space orbits ("strange attractor") in terms of the fractal dimension, the metric entropy, and the spectrum of Lyapunov exponents (the so-called "metric approach") and also in terms of the properties of periodic orbits embedded in the strange attractor (the so-called "topological approach") [see, for example, Mindlin et al. (1990) and the references cited therein]. In this paper the correlation dimension of the attractor is used to estimate the number of dynamically relevant modes.

We apply this approach to the study of geomagnetic disturbances in the auroral zone. It is well known that the VGF reflect the dynamics of the processes on the Sun, in the interplanetary medium, in the magnetosphere, and in the ionosphere. These processes have different physical origins and different time scales. Due to physical and geometric influences, the maximal "non-smooth" effects of these processes are observed in the ionosphere of the auroral zone and the polar cap where the evolution of VGF seems very complicated. (The auroral zone forms a ring around the geomagnetic pole and covers the interval of latitudes from 65 to 75 degrees. The polar cap is the area bounded by the poleward edge of the auroral zone and both the Northern and the Southern hemispheres have an auroral zone and a polar cap.) Let us note two peculiarities of the VGF. First, VGF have several natural time scales (some minutes, a day, a year, and the solar activity cycle) and the application of the arguments discussed above for different time scales can produce different dynamical systems. (In other words a different number of degrees of freedom can be important for different time scales.) Second, if one considers the VGF for a fixed time scale, then the control parameters of the underlying dynamical system (if such a system exists) can vary adiabatically with time.

The analysis of geomagnetic disturbances in the auroral zone is one of the tools for studying the solar wind-magnetospheric interaction. There are two possible objects for study. First, one can study the dynamics of individual magnetic storms. In this case, the temporal series of auroral geomagnetic indices, such as AE or AL, must be used. When the typical temporal interval of a magnetic storm is not more than a few hours, then 1-minute sampling intervals of 2.5-minutes data are used for the analysis. According to previous research (Vassiliadis et al., 1990; Roberts et al., 1991; Roberts, 1991; Shan et al., 1991a,b), the dynamics of the solar wind-magnetospheric system, which is responsible for an individual magnetic storm, lies on an attractor of dimension about 4 over a large range of activity levels. We will study another phenomena: the temporal variation of magnetospheric disturbances. The hourly values of the H component of the geomagnetic field in the auroral zone can be used in this case.

The geomagnetic field variations recorded by one station do not present complete information about the magnetospheric substorm, if for example the observatory is located in the dayside sector in the path of the substorm. Global modes are however rather well statistically discriminated by records of only one auroral station if data over a reasonably long time interval are analyzed. Another peculiarity of VGF analysis is the following. There are four main manifestations of magnetic field variations in the auroral zone (Nishida, 1978), namely, S_q-variation (with magnitude about 10 – 30 nT), DP2 variations (with magnitude < 40 nT), different pulsations (with magnitude < 50 nT), DP1 variations (with magnitude < 200 nT). The characteristic values of their magnitudes more or less depend on the latitude of the observation point, on the season, and so on. (There might be other constituents as well whose nature has not yet been determined.) It can be assumed that the dynamical system A modeling the geomagnetic field evolution inside the auroral zone (if such a system really exists) comprises two subsystems

$$A = B(\ell, t) * S \quad ,$$

where subsystem S describes global modes, which cover the entire auroral zone, and subsystem $B(\ell, t)$ is responsible for local (control parameter ℓ) and seasonal (control parameter t) effects. (When we say that the dynamical system A models the dynamics of the geomagnetic field, we suppose that the dynamic variables describe all essential modes. However, when the characteristics of the system A are restored by an observed time series of a finite length T, the degrees of freedom, connected with characteristic time scales $> T$, apparently can not be taken into account. In this case it is usually assumed that the system A depends on control parameters which might be considered constant in the time interval $< T$, but which, in general, vary for longer time intervals). It is generally accepted that the influence of the ionosphere on geomagnetic field fluctuations in the auroral zone depends on the season. We propose that the time parameter t is constant during the season, though a variation occurs from season to season with the most important changes after transition between autumn-winter and spring-summer periods. As the characteristic amplitudes of fluctuations, generated by a subsystem B, are less than the characteristic amplitudes of some of the global modes disturbances, we further propose that the analysis of the VGF data from individual stations allows us to obtain information not only about the complex system A, but the subsystem S as well.

NOTATION AND ALGORITHM

The algorithm for the estimation of the dimension of an attractor (the so-called correlation dimension) on the basis of an experimental series was proposed by Grassberger and Procaccia (1983) [GP algorithm]. We review briefly the calculation procedure and introduce the notation.

Let $u_i, i = 1, ..., N$ be values of a scalar observable $u(t)$ measured during the time interval $0 < t < T$ with a regular time increment, $dt = 1$. Choosing an integer $m > 0$, we reconstruct the set of m-dimensional vectors

$$X_i = \left(u_i, u_{i+1}, u_{i+2}, ..., u_{i+(m-1)}\right), \quad i = 1, ..., N - (m-1)$$

in the so-called embedding space (m is called the embedding dimension). Let $R(r)$ be the number of pairs $\{i, j\}$, such that the distance $S(X_i, X_j)$ between vectors X_i and X_j is less than r for a given $r > 0$. This function depends in principle on the embedding dimension m. If for small r

$$\log R(r) = b + d_m \log r$$

and $d_m = d$ does not depend on m for $m > m_0$, then d is the correlation dimension of the attractor. [When the GP algorithm gives us a finite value for the correlation dimension, it may not indicate deterministic chaos. For small data sets, fractional Brownian motion may show an anomalous scaling which can be interpreted as a finite dimension (Osborne and Provenzale, 1989). We assume that the solar wind-magnetospheric interaction is not such system. See also Roberts (1991).] In this way, a plot of $\log R(r)$ against $\log r$ must be linear for small r and becomes distorted at larger r because of non-linear effects. (Note that r cannot be too small. A lower limit to r is defined by two factors: the accuracy of the measurements and poor statistics as the result of the finite length of a time series.)

An important restriction of the GP algorithm is defined by the following circumstances (Ruelle, 1990): on one hand, the time increment dt can not be very small because the neighboring values of the time series under consideration may be strongly correlated in this case; on the other hand, the series of data have to be so long that

$$d < 2 \log N$$

where d is the correlation dimension and N is the number of values of a time series under consideration. To reduce the effect of the correlation of the values u_i, a slightly modified algorithm for the construction of the vectors X_i is used (Packard et al., 1980; Takens, 1981). Specifically, vectors X_i are constructed as follows:

$$X_i = \left(u_i, u_{i+l}, u_{i+2l}, ..., u_{i+\ell(m-1)}\right), \quad i = 1, 2, ...$$

The value of the parameter ℓ (delay time) depends on the series under consideration. The selection of the parameter ℓ was discussed in a number of papers (see, for example, Roberts [1991] and references cited therein). We shall select the parameter ℓ of the order of the correlation time for the temporal series under consideration and verify that the estimation of the correlation dimension does not depend on small variations of this parameter.

DESCRIPTION OF THE INITIAL DATA SETS

We analyzed variations of the H component of geomagnetic field measured at Fort Churchill (FCC), Barrow (BRW), Dixon Island (DIK), Great Whale River (GWC), Leirvogur (LRV), Tixie Bay (TIK), Abisco (ABK), and College (CMO) observatories. We used the hourly values of 1979 and 1980 which are routinely calculated by magnetic observatories on the basis of magnetograms. The results are similar for both of these years and we will discuss only the results of the estimation of the correlation dimension on the basis data of 1979. We discussed in the introduction that a regular S_q variation occurs in the time series considered. Since the origin of S_q is geometrical, it seems appropriate to exclude it from this discussion. However, owing to the rather small amplitude of S_q in the auroral zone against other fluctuations its influence on the correlation dimension proves not to exceed the limits of accuracy of this estimation (see Figure 1) and we will not exclude S_q variations from analyzed data. [See also Theiler et al. (1991) for a discussion of the effect of "pre-whitening" data.]

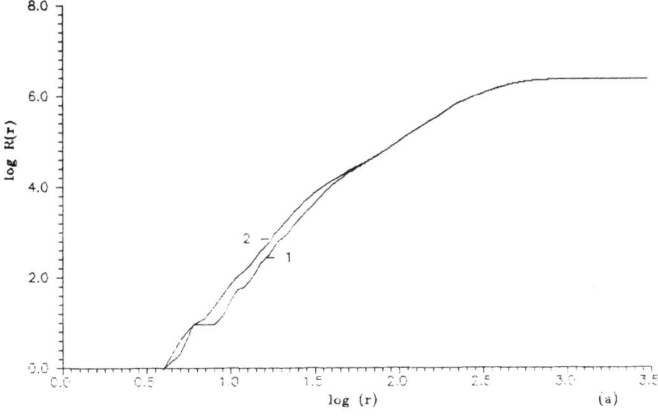

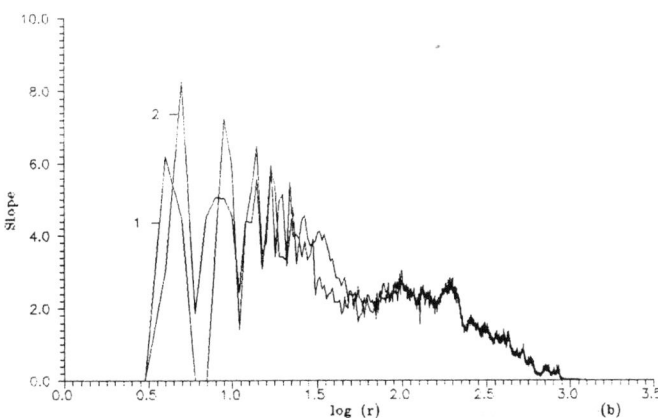

Fig. 1. (a) Plots of $\log R(r)$ versus $\log r$ for the H component of the VGF, observatory Tixie Bay, January–March of 1979, delay time = 4 hr, $m = 11$, 1 — for initial data, 2 — for data after deduction of S_q variation; (b) Plots of the pairwise logarithmic slopes of the curves in Figure 1a. The zero slope regions for small r are due to the 1 nT quantization of data.

RESULTS AND DISCUSSION

Figure 2a shows plots of $\log R(r)$ as a function of $\log(r)$ for Tixie Bay observatory data for January–March, 1979, for values of the embedding dimension of $m = 5, 7, 9, 10, 11, 12$ and a delay time $\ell = 4$ hr. Plots of the slope variations of these curves with $\log(r)$ are presented in Figure 2b. It is easily recognized that for $m > 9$ the values of d may be regarded as asymptotic. Typical variations of $\log R(r)$ with $\log(r)$ for four seasons are presented in Figure 3 (here and further we present the function $R(r)$ only for $r > 9$ nT, since values of this function for smaller r are strongly distorted owing to poor statistics). One can see that $R(r)$ contains two scaling regions for the autumn-winter period (see Figure 3) so that it has a slope d_1 in the range $\log r_1 - \log r_2$ and a slope d_2 in the range $\log r_3 - \log r_4$. Such behavior of $R(r)$ may mean—see Eckmann and Ruelle [1985]—that a dynamical system A consists of two sub-systems B and S ($A = B * S$), so that an observable

$$u_A(t) = u_S(t) + u_B(t) \quad ,$$

and the amplitude of a signal $u_B(t)$ generated by the subsystem B is much less than the amplitude of a signal $u_S(t)$ generated by the subsystem S (we suggest that subsystems B and S are independent or that subsystem S weakly influences subsystem B).

The disappearance of the "knee" of the function $\log R(r)$ versus $\log r$ in the spring-summer period means that the amplitude of the signal u_B increases and becomes comparable to the amplitude of the signal u_S. Apparently, it also accounts for the general increase of the correlation dimension during the spring-summer period (see Table 1). The dependence of the correlation dimensions of both subsystems on the latitude is presented in Table 2. The "knee" in the plot of the function $R(r)$ gradually disappears as one approaches the middle of the auroral zone. It is absent for Great Whale River (GWC) station which is located near the middle of the auroral zone. Possibly, this means that the influence of the subsystem B on the chaotic behavior of the VGF increases in the middle area of the auroral zone. Two subsystems are discriminated again near the poleward edge of the auroral zone. The value of correlation dimension d gives us the lower boundary of the number of degrees of freedom which are responsible for the chaotic behavior of the system. In other words, the smallest integer $k > d$ gives us the minimal number of ordinary first order differential equations for modeling the evolution of the system. This suggests that an additional degree of freedom could be switched on during the spring-summer period in comparison with the autumn-winter period.

We identify the subsystem S as the system which is responsible for the dynamics of time variation of magnetospheric disturbances. It should be noted in comparison that Roberts (1991) showed that when the time series of 2.5 min auroral index is filtered giving a time base of about 25 min, the dimension determined by GP algorithm is near 2. We did not use such a filter and the value of correlation dimension of subsystem S can not be explained by such an effect.

Acknowledgements. The authors are grateful to WDC A for Solar-Terrestrial Physics for kind opportunity to use geomagnetic database on CD-ROM, Dr. D.A. Roberts for valuable comments and Dr. D. Bilitza for assistance. We are grateful to Prof. W. Newman for his kind cooperation in preparing the final version of the manuscript.

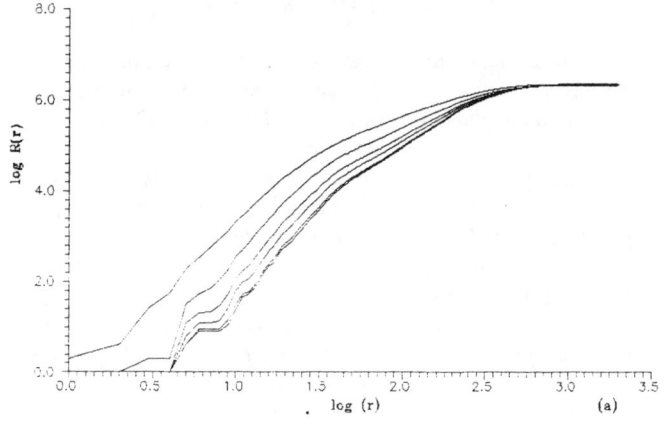

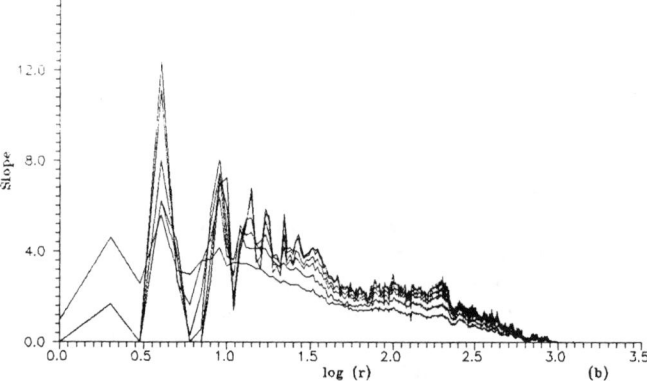

Fig. 2. (a) Plots of log $R(r)$ versus log r for the H component of the VGF, observatory Tixie Bay, January–March of 1979, delay time = 4 hr, m = 5, 7, 9, 10, 11, 12 with m increasing from top to bottom; (b) Plots of the pairwise logarithmic slopes of the curves in Figure 2a, m increases from bottom to top in the middle part of graphs.

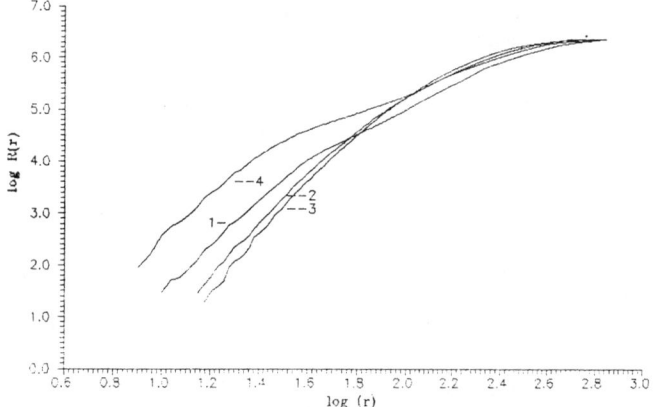

Fig. 3. Plots of log $R(r)$ versus log r for four seasons of the 1979 year for the H component of the VGF measured at observatory Tixie Bay: 1— January–March, 2 — April–June, 3 — July–September, 4 — October–December. Delay time = 4 hr, m = 11. (The points for r < 9 nT are deleted).

international code of observatory	quarter of the year 1979	correlation dimension subsystems B * S	subsystem S
C M O	1	5.07	2.39
	2	5.56	—
	3	5.83	—
	4	5.02	1.59
T I K	1	4.56	2.25
	2	5.16	—
	3	5.76	—
	4	4.34	1.60
L R V	1	4.85	—
	2	5.27	—
	3	6.61	—
	4	4.70	1.70
B R W	1	4.83	?
	2	5.61	—
	3	6.64	—
	4	4.37	2.59

TABLE 1. Estimations of the correlation dimension on the basis of hourly values of the H component of geomagnetic variations.

international code of observatory	correlation dimension subsystems B * S	subsystem S
C M O	5.02 ?	1.59
A B K	4.36	1.60
T I K	4.34	1.60
L R V	4.70	1.70
G W C	4.80	—
D I K	?	2.90
B R W	4.37	2.59
F C C	5.73 ?	2.74

TABLE 2. Estimation of the correlation dimension on the basis of hourly values of the H component of geomagnetic variations for the fourth quarter of 1979.

References

Anderson, P.W., K.J. Arrow, and D. Pine (eds.), *The Economy as an Evolving Complex System,* 354 pp., Redwood City, California: Addison-Wesley, 1988.

Eckmann, J.P., and D. Ruelle, Ergodic theory of chaos and strange attractors, *J. Math. Phys.*, 57, 617–625, 1985.

Ghil, M., R. Benzi, and G. Parisi (eds.), Turbulence and Predictability in *Geophysical Fluid Dynamics and Climate Dynamics*, 289 pp., Amsterdam: North-Holland, 1985.

Grassberger, P., Do climate attractor exist? *Nature*, 323, 609–612, 1986.

Grassberger, P., and I. Procaccia, Measuring the strangeness of strange attractors, *Physica D*, 9, 189–208, 1983.

Lorenz, E.M., Deterministic nonperiodic flow, *J. Atmos. Sci.*, 20, 130–141, 1963.

Mandelbrot, B.B., *The Fractal Geometry of Nature*, 235 pp, San Francisco: Freeman, 1982.

Mindlin, G.B, X.-J. Hou, H.G. Solari, R. Gilmore, and N.B. Tufillaro, Classification of strange attractors by integers, *Phys. Rev. Lett.*, 64, 2350–2353, 1990.

Moon, F., *Chaotic Vibrations: An Introduction for Applied Scientists and Engineers*, 267 pp., Wiley-Interscience Publication, 1987.

Nicolis, C., and G. Nicolis, Is there a climatic attractor? *Nature*, 311, 529–532, 1984.

Nicolis, C., and G. Nicolis, Evidence for climatic attractors,*Nature*, 326, 523-525, 1987.

Nishida, A., *Geomagnetic Diagnosis of the Magnetosphere*, 197 pp., Springer-Verlag, 1978.

Osborne, A.R., and A. Provenzale, Finite correlation dimension for stochastic systems with power-law spectra, *Physica D*, 35, 357–369, 1989.

Packard, N.H., J.P. Crutchfield, J.D. Farmer, and R.S. Shaw, Geometry from a time series, *Phys. Rev. Lett.*, 45, 712–716, 1980.

Rapp, P.E., I.D. Zimmerman, A.M. Albano, G.C. Deguziman, N.N. Grenbaum, and T.R. Bashor, Experimental studies of chaotic neural behavior: cellular activity and electroencephalographic signs, in *Nonlinear Oscillations in Chemistry and Biology*, edited by H.G. Othmer, pp. 154-167, New York: Springer, 1987.

Roberts, D.A., Is there a strange attractor in the magnetosphere?, *J. Geophys. Res.*, 96, 16,031–16,040, 1991.

Roberts, D.A., D.N. Baker, A.J. Klimas, and L.F. Bargatze, Indications of low dimensionality in magnetospheric dynamics, *Geophys. Res. Lett.*, 18, 151–155, 1991.

Ruelle, D, Diagnosis of dynamical systems with fluctuating parameters, *Proc. R. Soc. Lond.*, A413, 5–12, 1987.

Ruelle, D., Deterministic chaos: the science and the fiction, *Proc. R. Soc. Lond.*, A427, 241–253, 1990.

Shan, L.-H., C.K. Goertz, and R.A. Smith, On the embedding-dimension analysis of AE and AL time series, *Geophys. Res. Lett.*, 18, 1647–1651, 1991a.

Shan, L.-H., P. Hansen, C.K. Goertz, and R.A. Smith, Chaotic appearance of the AE index, *Geophys. Res. Lett.*, 18, 147–152, 1991b.

Takens, F., Detecting strange attractors in turbulence, in *Dynamical Systems and Turbulence*, pp. 366–381, Berlin: Springer, 1981.

Theiler, J., B. Galdrikian, A. Longtin, S. Eubank, and J.D. Farmer, Using surrogate data to detect nonlinearity in time series. Los Alamos National Laboratory Rep. LA–UR–91–2615, 56 pp., 1991.

Tsonis, A.A, and J.B. Elsner, The weather attractor on very short time scales. *Nature*, 333, 545–547, 1988.

Vassiliadis, D.V., A.S. Sharma, T.E. Eastman, and D. Papadopoulos, Low–dimensional chaos in magnetospheric activity from AE time series, *Geophys. Res. Lett.*, 17, 1841–1845, 1990.

Yu. Tyupkin and A. Feldstein, Geophysical Center, Molodezhnaya 3, Moscow, 117296, Russia